SPACE SCIENCE, EXPLORATION AND POLICIES

NEAR-EARTH OBJECTS

IDENTIFYING AND MITIGATING POTENTIAL THREATS FROM SPACE

SPACE SCIENCE, EXPLORATION AND POLICIES

SPACE SCIENCE, EXPLORATION AND POLICIES

NEAR-EARTH OBJECTS

IDENTIFYING AND MITIGATING POTENTIAL THREATS FROM SPACE

LEON SINCLAIR
EDITOR

nova publishers
New York

NOTICE TO THE READER

The Publisher has taken reasonable care in the preparation of this book, but makes no expressed or implied warranty of any kind and assumes no responsibility for any errors or omissions. No liability is assumed for incidental or consequential damages in connection with or arising out of information contained in this book. The Publisher shall not be liable for any special, consequential, or exemplary damages resulting, in whole or in part, from the readers' use of, or reliance upon, this material. Any parts of this book based on government reports are so indicated and copyright is claimed for those parts to the extent applicable to compilations of such works.

Independent verification should be sought for any data, advice or recommendations contained in this book. In addition, no responsibility is assumed by the publisher for any injury and/or damage to persons or property arising from any methods, products, instructions, ideas or otherwise contained in this publication.

This publication is designed to provide accurate and authoritative information with regard to the subject matter covered herein. It is sold with the clear understanding that the Publisher is not engaged in rendering legal or any other professional services. If legal or any other expert assistance is required, the services of a competent person should be sought. FROM A DECLARATION OF PARTICIPANTS JOINTLY ADOPTED BY A COMMITTEE OF THE AMERICAN BAR ASSOCIATION AND A COMMITTEE OF PUBLISHERS.

Additional color graphics may be available in the e-book version of this book.

Library of Congress Cataloging-in-Publication Data

ISBN: 978-1-63484-067-5

Published by Nova Science Publishers, Inc. † New York

CONTENTS

PREFACE

Scientists classify comets and asteroids that pass within 28 million miles of Earth's orbit as near-Earth objects (NEOs). Asteroids that collide and break into smaller fragments are the source of most NEOs, and the resulting fragments bombard the Earth at the rate of more than 100 tons a day. Although the vast majority of NEOs that enter Earth's atmosphere disintegrate before reaching the surface, those larger than 100 meters (328 feet) may survive the descent and cause destruction in and around their impact sites. Furthermore, even smaller objects that disintegrate before reaching Earth's surface can cause significant damage. This book examines NASA's NEO Program and assesses the Agency's progress toward meeting statutory and other Program goals. Specifically, it reviews NASA's allocation and use of resources and plans for the future of the Program.

In: Near-Earth Objects
Editor: Leon Sinclair
ISBN: 978-1-63484-067-5

Chapter 1

NASA's Efforts to Identify Near-Earth Objects and Mitigate Hazards*

NASA Office of Inspector General

Acronyms

ATLAS	Asteroid Terrestrial-impact Last Alert System
DARPA	Defense Advanced Research Projects Agency
FY	Fiscal Year
JPL	Jet Propulsion Laboratory
KaBOOM	Ka-Band Objects Observation and Monitoring Project
LINEAR	Lincoln Near-Earth Asteroid Research
LSST	Large Synoptic Survey Telescope
NEA	Near-Earth Asteroid
NEO	Near-Earth Object
NEOCam	NEO Camera
NPD	NASA Policy Directive
NPR	NASA Procedural Requirements
NRC	National Research Council
NSF	National Science Foundation

* This is an edited, reformatted and augmented version of a report, IG-14-030, issued by the National Aeronautics and Space Administration's Office of Inspector General, September 15, 2014.

OIG	Office of Inspector General
OSIRIS-Rex	Origins-Spectral Interpretation-Resource Identification-Security - Regolith Explorer Project
Pan-STARRS	Panoramic Survey Telescope and Rapid Response System 1 and 2
PHO	Potentially Hazardous Object
SST	Space Surveillance Telescope
WISE	Wide-field Infrared Survey Explorer

OVERVIEW

The Issue

Scientists classify comets and asteroids that pass within 28 million miles of Earth's orbit as near-Earth objects (NEOs). Asteroids that collide and break into smaller fragments are the source of most NEOs, and the resulting fragments bombard the Earth at the rate of more than 100 tons a day. Although the vast majority of NEOs that enter Earth's atmosphere disintegrate before reaching the surface, those larger than 100 meters (328 feet) may survive the descent and cause destruction in and around their impact sites. Furthermore, even smaller objects that disintegrate before reaching Earth's surface can cause significant damage. For example, in February 2013 an 18-meter (59 foot) meteor exploded 14.5 miles above the city of Chelyabinsk, Russia, with the force of 30 atomic bombs, blowing out windows, destroying buildings, injuring more than 1,000 people, and raining down fragments along its trajectory (see Figure 1). Recent research suggests that Chelyabinsk-type events occur every 30 to 40 years, with a greater likelihood of impact in the ocean than over populated areas, while impacts from objects greater than a mile in diameter are predicted only once every several hundred thousand years.

In 1992, NASA began conducting scientific workshops and research into the identification, characterization, and tracking of NEOs, as well as into potential mitigation strategies. NASA reported NEOs with a diameter greater than 1 kilometer (0.62 miles) posed the greatest hazard to Earth and predicted a comprehensive survey could identify most NEOs of this size within a decade.[1] In 1994, the House Committee on Science, Space, and Technology requested NASA identify and catalogue within 10 years the orbital characteristics of all comets and asteroids greater than 1 kilometer in diameter and in an orbit around the Sun that crosses the orbit of the Earth.[2] Four years later, NASA established a NEO Program Office to coordinate these efforts.[3]

Figure 1. Photograph of Chelyabinsk Meteor.

In addition, the NASA Authorization Act of 2005 required the Agency to implement a "program to detect, track, catalogue, and characterize the physical characteristics of near-Earth objects equal to or greater than 140 meters in diameter" and established a goal of cataloging 90 percent of these objects by 2020.[4] However, even with a ten-fold increase in the NEO Program budget in the past 5 years – from $4 million in fiscal year (FY) 2009 to $40 million in FY 2014 – NASA estimates that it has identified only about 10 percent of all asteroids 140 meters and larger. Moreover, given its current pace and resources, the Agency has stated that it will not meet the goal of identifying 90 percent of such objects by 2020.

We initiated this review to examine NASA's NEO Program and assess the Agency's progress toward meeting statutory and other Program goals. Specifically, we reviewed NASA's allocation and use of resources and plans for the future of the Program. Details of the review's scope and methodology are in Appendix A.

Results

NASA has organized its NEO Program under a single Program Executive who manages a loosely structured conglomerate of research activities that are not well integrated and lack overarching Program oversight, objectives, and established milestones to track progress. In addition, NASA is undertaking NEO-related activities not managed by the Program and not sufficiently integrated into ongoing Program activities. Furthermore, NASA lacks formal agreements or procedures for NEO-related activities it conducts with other Federal agencies and foreign governments and has not taken advantage of possible partnership opportunities. Consequently, managers could not identify the level of resources required to adequately support the Program or explain how activities to which the NEO Program is contributing further Program goals. Even though the Program has discovered, categorized, and plotted the orbits of more than 11,000 NEOs since 1998, NASA will fall short of meeting the 2005 Authorization Act goal of finding 90 percent of NEOs larger than 140 meters in diameter by 2020. We believe the Program would be more efficient, effective, and transparent were it organized and managed in accordance with standard NASA research program requirements.

Lack of Structure and Limited Resources Hinders the NEO Program. Since creation of the NEO Program, the number of identified NEOs has increased from less than 500 in 1998 to over 11,000 as of July 2014. Most of this work occurred while the Program was relatively small – receiving only $4 million annually from FYs 2002 through 2009 and funding less than 20 individual efforts – and focused on identifying the largest NEOs.

However, with the directive to identify 90 percent of NEOs larger than 140 meters, substantially increased budget beginning in FY 2011, and additional projects, the NEO Program's existing structure and resources are inadequate to provide efficient, effective, and transparent program management.

In FY 2013, the Program Executive oversaw a budget of $20.5 million and 64 funding instruments that included grants, purchase orders, and contract task orders to observatories and other facilities. With the Program budget growing to $40 million in FY 2014, the number of funding instruments will likely also increase and with it the Program Executive's oversight responsibilities.

Source: NASA.

Figure 2. Arecibo Observatory.

In addition to limited personnel, the NEO Program lacks a plan with integrated milestones, defined objectives, and cost and schedule estimates to assist in tracking and attaining Program goals. To implement the goals of the 2005 Authorization Act, the Program provided funding to more ground-based observatories and obtained additional observation time at observatories such as the Panoramic Survey Telescope and Rapid Response System in Hawaii and the Arecibo Observatory in Puerto Rico (see Figure 2). In addition, the Program supports the work of NASA initiatives such as the Asteroid Redirect Mission and NEO Program personnel provide technical support for a Space Act Agreement with the B612 Foundation to assist in the development of a privately funded, space-based infrared telescope.[5] Despite this increased activity, NASA has not changed or improved the NEO Program's management structure and the Program has not established a plan to integrate the additional initiatives or track their contributions to attainment of NEO Program goals.

In our judgment, the N EO Program would be better equipped to meet its goals if the Program followed NASA's programmatic policies for research projects and if Program management established clear and verifiable requirements for processes pertaining to N EO detection, characterization, and mitigation.[6]

NASA Needs to Improve its Oversight of NEO Grants and Task Orders. NASA's controls for managing and overseeing costs associated with the NEO Program are inadequate to ensure proper accounting of Agency-funded grants and task orders. First, we found that contrary to effective internal control standards, the NEO Program Executive is responsible for or has significant input into all the primary elements of the award process, including overseeing, monitoring, and evaluating the progress of awards.[7] Having the Program Executive perform all these tasks increases the risk of error or fraud in the Program.

We also reviewed the 10 awards that exceeded $900,000 the NEO Program funded in FY 2013 and found the Program did not adequately oversee, monitor, or evaluate the progress of the work performed. For example, NASA recently awarded the Lincoln Near-Earth Asteroid Research Program a $6.5 million, 5-year grant via an interagency transfer in spite of a steep decline in its ability to provide NEO detection. In addition, we identified three instances in which NASA awarded grants when contracts would have been the more appropriate instrument for achieving Program goals. Although these grants supported some research, the majority of funds supported operations and maintenance on observatory facilities and therefore, pursuant to Federal law, NASA should have used a contract.[8] Furthermore, a contract would have provided greater visibility into awardee operations and ensured the level of funding and awardee performance was commensurate with requirements and deliverables.

Lack of Formal Agreements with Federal and International Partners Hampers NASA's Ability to Accomplish Program Goals. Although NASA has established two formal partnerships with domestic, nongovernmental research organizations and several informal partnerships, a lack of planning and resources has prevented the NEO Program from developing additional partnerships that could help achieve Program goals. For example, establishing formal partnerships with the Department of Defense and the National Science Foundation could give the Program access to additional Earth-based telescopes, thereby increasing its ability to detect, track, and characterize a greater number of NEOs. Moreover, the NEO Program has not taken advantage of experience gained by other NASA programs, such as the International Space Station, that have formed international partnerships. Such expanded partnerships could accelerate discovery of NEOs, aid in development of global mitigation strategies, and help ensure that the burden of mitigating NEO threats does not disproportionately fall on NASA and the United States.

Management Action

To improve NASA's efforts to discover, characterize, and catalog NEOs and develop mitigation strategies, we recommended the Associate Administrator for the Science Mission Directorate develop a formal NEO Program with a strategic plan, integrated master schedule, and cost estimates. We also recommended the Associate Administrator direct the NEO Program Executive to develop and implement requirements, procedures, and internal controls to address deficiencies; perform a full-time equivalent analysis to determine the staff required to manage, oversee, and administer the Program; develop a plan to establish formal partnerships with domestic and international agencies to leverage resources and complementary technologies; and establish a systematic oversight process pursuant to which NASA-funded observatories are required to coordinate to avoid duplication of effort.

In response to a draft of this report, the Associate Administrator concurred with our recommendations and agreed to establish a formal NEO Program in accordance with NASA guidance. He also stated the Program will conduct a full-time equivalent analysis, develop a plan to establish formal partnerships, and establish a coordination plan between observatories – all of which he promised will be documented in the NEO Program Plan.

We consider NASA's planned actions responsive and will close the recommendations upon verification of their completion. We also reviewed management's comments regarding the technical accuracy of the draft and made changes as appropriate. Management's full response to the draft report is reprinted in Appendix D.

INTRODUCTION

Background

Scientists classify comets and asteroids that pass within 28 million miles of Earth's orbit as near-Earth objects (NEOs).[9] Composed of rock, ice, and organic compounds, comets are thought to originate from an area beyond the orbit of Pluto but occasionally have orbits that bring them closer to the Sun where they heat up and may display a visible atmosphere and tail due to the effects of solar wind and radiation. Asteroids are primarily rocky or metallic bodies that orbit the Sun, with most residing in the main asteroid belt between Mars and Jupiter. Asteroids that collide and break into smaller fragments are the source of most NEOs. Fragments of comets or asteroids that enter the Earth's atmosphere are known as meteors. Most meteors are small and vaporize in the Earth's atmosphere as "shooting stars" before reaching the planet's surface. A meteor that reaches the Earth's surface is known as a meteorite.

Every day more than 100 tons of meteors bombard the Earth. Although the vast majority of these objects disintegrate before reaching the planet's surface, objects larger than 100 meters (328 feet) may survive the descent, hit the ground, and cause destruction in and around the impact site. Moreover, even smaller objects that enter the Earth's atmosphere and disintegrate before reaching the surface can cause significant damage. For example, on February 15, 2013, an 18-meter (59 foot) meteor exploded 14.5 miles above the city of Chelyabinsk, Russia, with the force of 30 atomic bombs, blowing out windows, destroying buildings, injuring more than 1,000 people, and raining down fragments along its trajectory.[10] Recent research suggests that Chelyabinsk-type events occur every 30 to 40 years with a greater likelihood of impact in the ocean than over populated areas.

Although the probability of a meteor 1 kilometer (3,280 feet) or larger striking the Earth is extremely remote, the consequences of such an impact would be severe. On May 31, 2013, a massive asteroid 2.7 kilometers (1.6 miles) in diameter passed within 5.8 million kilometers (3.6 million miles) of Earth or about 15 times the distance from the Earth to the Moon. Had an object this size struck the Earth, the resulting debris would likely have contaminated the Earth's atmosphere, causing partial obstruction of sunlight, acid rain, and firestorms. In a dramatic example of this phenomenon, scientists believe that a 10-kilometer wide (6.2 mile) meteorite struck Mexico's Yucatan Peninsula about 66 million years ago, perhaps contributing to the extinction of the dinosaurs. Table 1 summarizes impact frequencies based on meteor size.

Legislative Directives and National Research Council Findings

NASA conducts scientific workshops and performs research into identifying and characterizing the physical and orbital properties of and tracking NEOs.[11] The Agency also funds and conducts research into mitigating potentially hazardous NEOs. In 1992, NASA issued a report concluding NEOs with a diameter larger than 1 kilometer posed the greatest hazard to Earth and posited that a comprehensive survey could identify most NEOs of this size within a decade.[12] In 1994, the House Committee on Science, Space, and Technology requested NASA identify and catalogue within 10 years the orbital characteristics of all comets and asteroids that are greater than 1 kilometer in diameter and are in an orbit around the Sun that crosses the orbit of the Earth.[13] Thereafter, NASA's Authorization Act of 2005 required the Agency to implement a "program to detect, track, catalog, and characterize the physical characteristics of near-Earth objects equal to or greater than 140 meters in diameter in order to assess the threat of such near-Earth objects to the Earth" and set a goal of cataloguing 90 percent of these objects by 2020.[14] The law also amended the National Aeronautics and Space Act of 1958 and additionally required NASA use its skills to "provide warning and mitigation of the potential hazard of such near-Earth objects to the Earth."[15]

Table 1. Approximate Impact Frequencies and Consequences from Near-Earth Objects

Type of Event	Diameter of Object (meters)	Impact Energy (megatons)[a]	Average Impact Interval (years)
High al titude break-up	< 30	<5	1–50
Tunguska-like event	> 30	>5	250–500
Regional event	> 140	~150	5,000
Large sub-global event	> 300	~2,000	25,000
Low global effect	> 600	~30,000	70,000
Medium global effect	> 1,000	>100,000	1 million
High gl obal effect	> 5,000	> 10 million	6 million
Exti ncti on-cl ass event	> 10,000	>100 million	100 million

[a] A megaton is a unit of measurement describing the amount of energy released from an explosion. One megaton equals the amount of energy released from one million tons of TNT.

Source: NASA.

In 2010, the National Research Council (NRC) issued a report finding NASA would not meet the goal of identifying 90 percent of NEOs larger than 140 meters by 2020 due to insufficient resources.[16] The NRC identified two approaches to reach the goal by 2030 depending on the priority policymakers attached to the issue. Specifically, if finishing the survey as soon as possible was most important, the NRC recommended using a space-based telescope in concert with observations from suitable ground-based telescopes. If conserving costs was a higher priority, the NRC suggested using only large ground-based telescopes.

The NRC report recommended NASA be directed to monitor objects as small as 30 to 50 meters in diameter. In addition, the NRC suggested the

United States organize an international entity to develop a detailed plan for addressing NEO hazards and NASA and the National Science Foundation (NSF) support a vigorous program of observation at the Arecibo Observatory in Puerto Rico (Arecibo) and Goldstone Deep Space Communications Complex in California (Goldstone), both of which play an important role in determining the orbits and characterizing the properties of NEOs.

NASA has surveyed about 95 percent of the known population of NEOs 1 kilometer or larger and significantly increased efforts to locate and characterize NEOs between 140 meters and 1 kilometer. As of July 2014, approximately 11,230 NEOs have been identified of which 862 have diameters of approximately 1 kilometer or larger. In addition, NASA estimates that it has identified approximately 10 percent of all asteroids 140 meters and larger. Moreover, NASA has classified 1,492 NEOs as potentially hazardous objects (PHOs), meaning they have orbits predicted to come within roughly 4.6 million miles of Earth's orbit and exceed about 150 meters (500 feet) in diameter. Figure 3 illustrates the quantity of known asteroids by size as of July 2014.

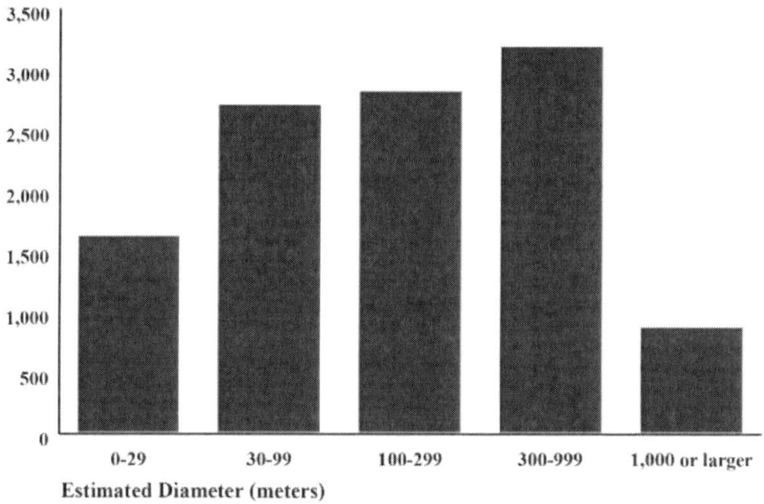

Note: Comet size is unknown; therefore, the 94 known near-Earth comets are not included. Source: NASA.

Figure 3. Sizes of Near-Earth Asteroids Discovered.

NASA's NEO Program

In 1998, NASA established a NEO Program Office in the Science Mission Directorate's Planetary Science Division to coordinate its NEO-related efforts.[17] The Program Executive sits at NASA Headquarters and is responsible for overall execution of the NEO Program and oversight of related grants and contracts. The Executive is the only NEO Program employee at Headquarters. The other major parts of the Program are the Minor Planet Center in Cambridge, Massachusetts, and the NEO Program Office at the Jet Propulsion Laboratory (JPL).

NASA funds the Minor Planet Center through a 5-year, $6.3 million grant. Hosted by the Harvard-Smithsonian Center for Astrophysics, the Center is an international clearinghouse for small body observational data under the authority of the International Astronomical Union. A director, two investigators, and three information technology employees staff the Center.

The NEO Program Office receives, refines, and publishes data on the orbits and impact likelihood of objects reported to the Minor Planet Center. NASA funds the Office through a 14-year, $8.9 million task order issued pursuant to NASA's larger contract with the California Institute of Technology to operate JPL. The Office has six employees – a manager, four research scientists, and an information technology specialist.[18]

Since 1998, NASA has spent about $100 million in NEO-related efforts using funds from the Science Mission Directorate's Planetary Science Research Program. Over the past 5 years, the annual budget has grown from approximately $4 million in fiscal year (FY) 2009, to $20 million in FYs 2011 through 2013, to $40 million in FY 2014.[19]

NEO Program Operations

NEO Program operations consist of three major elements: detection and notification, characterization and cataloging, and mitigation (see Appendix B for more detail on the Program assets that support these elements).

- *Detection and Notification*. The ability to detect a NEO is dependent on the object's distance from the Earth; its size and location relative to the Sun; and how well light reflects from its surface, also known as its "albedo." It is extremely difficult, if not impossible, to detect a NEO that is near or, as observed from Earth, positioned in front of the Sun. The NEO Program provides funds to three survey teams that operate five ground-based telescopes to detect NEOs – the Catalina Sky Survey (Catalina), Lincoln Near-Earth Asteroid Research (LINEAR), and Panoramic Survey Telescope and Rapid Response System 1 and 2 (Pan-STARRS). Together these teams have reported approximately 73 percent of all NEO observations recorded by the Minor Planet Center. To supplement these Earth-based surveys, in September 2013 NASA reactivated and reprogrammed its Wide-field Infrared Survey Explorer (WISE) telescope to search for NEOs.[20] Renamed NEOWISE, this space-based infrared telescope is particularly useful for discovering dark-colored asteroids that are extremely difficult to detect with ground-based telescopes.

- *Characterization and Cataloging.* To determine threat potential and plan any necessary mitigation efforts, scientists must characterize the composition, size, shape, and orbital properties of a NEO. For example, a NEO composed of iron is denser than one composed of rock and therefore more likely to survive the descent through Earth's atmosphere. Moreover, to plan effective mitigation efforts authorities need to know a NEO's path and likely impact site. The NEO Program provides funding to Arecibo and Goldstone to assist in the effort to characterize NEOs and the Minor Planet Center and the NEO Program Office maintain a catalog of the characteristics of specific NEOs.
- *Mitigation.* M itigati on is the means of defending Earth and its inhabitants from the effects of NEO impacts. M iti gati on may take the form of civil defense efforts such as emergency evacuations or efforts to deflect the trajectory of an object such as "slow-push" or "slow-pull," kinetic impact, or nuclear explosions in space.[21] The NEO Program devotes about $1 million annually or 7 percent of its overall funding to the study of mitigation strategies.

The NEO Program implemented the process illustrated in Figure 4 to meet the goal of identifying, characterizing, and mitigating NEOs of 140 meters or larger.

Generally, professional and amateur astronomers report their observations of possible NEOs to the Minor Planet Center, which acts as a repository for this information (see Appendix C for a list of NASA-funded and other observatories). The Center collects, catalogs, and disseminates for further analysis information about NEOs so that scientists can characterize the objects and authorities can plan any necessary mitigation efforts. Accordingly, the Center is a critical element in NASA's NEO Program and vital to the Program's success. By comparing reported objects to a database of known orbits, the Center's automated computer system verifies and validates reported objects to determine if they represent a new discovery. In the case of a new discovery, the Center calculates and publishes an initial orbit so that observatories throughout the world can search for and confirm the object's existence. If a NEO is determined to be an unidentified PHO with a potential for impact, NEO Program Office personnel use the Sentry System, an automated collision monitoring system, to continually scan the most current asteroid catalog and conduct a more detailed and refined analysis. The Program attempts to determine the precise orbit of NEOs, publishes results of its research, and requests follow-up observations from the scientific community. Previously reported data is continually refined to determine whether a PHO has the potential to affect the Earth in the next 100 years. If the Program Office determines there is no potential for harm, it works with the Minor Planet Center to refine and reconcile any differences in orbital calculations.

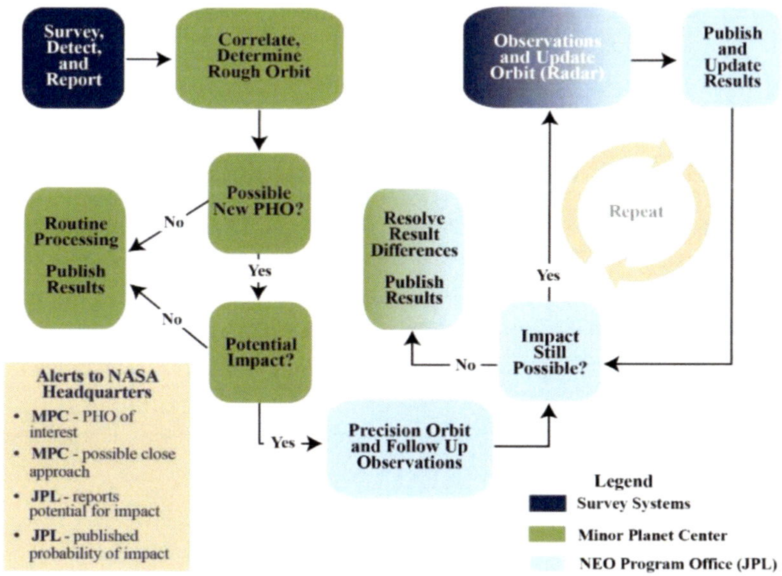

Note: Minor Planet Center (M PC). Source: NASA.

Figure 4. NEO Program Operation Process.

An example of this process is Asteroid 2014 AA, a 2–3 meter (7–10 foot) asteroid first detected by a team at California's Catalina Sky Survey Lunar and Planetary Laboratory between 1:18 a.m. and 1:46 a.m. Eastern Standard Time on January 1, 2014. The Catalina team quickly determined a potential impact trajectory of the asteroid and reported their findings to the Minor Planet Center. The Center and the NEO Program Office each performed independent orbit calculations and correctly predicted Asteroid 2014 AA would enter the Earth's atmosphere and travel along an arc extending from Central America to East Africa, with impact most likely to occur off the coast of West Africa at approximately 9:00 p.m. Eastern Standard Time that same day.

Related NASA and Federal Government Initiatives. The NEO Program provides funding to several other related NASA projects, including the Ka-Band Objects Observation and Monitoring Project (KaBOOM), the Asteroid Terrestrial-impact Last Alert System (ATLAS), and the NEO Camera (NEOCam) infrared space telescope.[22] In addition, Program personnel contribute to NASA's Asteroid Redirect Mission and Origins-Spectral Interpretation-Resource Identification-Security-Regolith Explorer Project (OSIRIS-Rex).[23] Other Federal agencies are also involved in NEO-related initiatives, including the Defense Advanced Research Projects Agency (DARPA), the NSF, and the Department of Energy. See Appendix B for more detail on these initiatives.

Objectives

We initiated this review to examine NASA's NEO Program and assess the Agency's progress toward meeting statutory and other Program goals. Specifically, we reviewed NASA's allocation and use of resources and plans for the Program's future. See Appendix A for details of the review's scope and methodology, our review of internal controls, and a list of prior coverage.

RESULTS

Existing Neo Program Management Structure Not Commensurate with Increased Resources and Expanded Responsibilities

We found that NASA has organized its NEO Program under a single Program Executive who manages a loosely structured conglomerate of research activities that are not well integrated and lack an overarching framework with Program oversight, objectives, and established milestones to track progress. In addition, NASA is undertaking NEO-related activities not managed by the Program and not sufficiently integrated into Program activities. Furthermore, NASA lacks formal agreements or procedures for NEO-related activities it conducts with other Federal agencies and foreign governments and has not taken advantage of some partnership opportunities. Consequently, Program managers could not identify the level of resources required to support the Program adequately or explain how some activities to which the NEO Program is contributing further Program goals. Even though the Program has discovered, categorized, and plotted the orbits of more than 11,000 NEOs since 1998, NASA will fall short of meeting the 2005 Authorization Act goal of finding 90 percent of NEOs larger than 140 meters in diameter by 2020. In sum, we believe the Program would be more efficient, effective, and transparent were it organized and managed in accordance with standard NASA research program requirements.

Lack of Structure and Limited Resources Hinders the NEO Program

Since implementation of the NEO Program, the number of identified NEOs has increased dramatically from less than 500 in 1998 to over 11,000 as of July 2014, including nearly 900 objects larger than 1 kilometer in diameter (see Figure 5). Most of this work occurred while the Program was relatively small – receiving only $4 million annually from FYs 2002 through 2009 and funding less than 20 individual efforts – and focused on identifying the largest NEOs. However, with the 2005 statutory directive to identify 90 percent of NEOs larger than 140 meters, a substantially increased budget beginning in FY 2011, and additional research projects, the existing programmatic structure and resources are not sufficient to provide for efficient, effective, and transparent program management.

NASA has placed overall Program responsibility in a single Program Executive at Headquarters who has no dedicated staff to assist with Program oversight. In FY 2013, the Program Executive oversaw a budget of $20.5 million and 64 funding instruments, including grants, purchase orders, and contract task orders to observatories and other facilities. More than half of these instruments exceeded $100,000, and nine involved more than $1 million in annual funding. With the Program budget growing to $40 million in FY 2014, the number of funding instruments will likely also increase, and with it the Program Executive's oversight responsibilities.

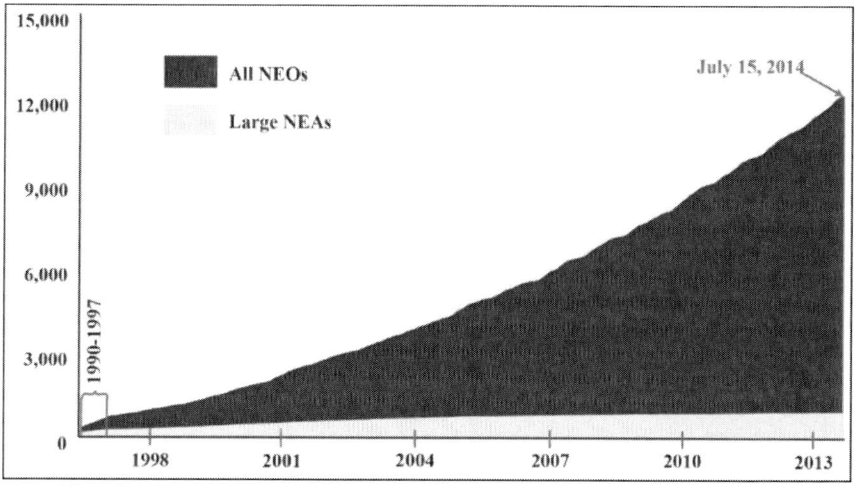

Note: Near-Earth asteroids (NEA). Comet size is unknown; therefore, 94 known near-Earth comets are only accounted for in the total population of all NEOs.
Source: NASA.

Figure 5. Number of Near-Earth Objects Discovered.

In addition to limited personnel, the NEO Program lacks a plan with integrated milestones, defined objectives, and cost and schedule estimates to assist in tracking and attaining Program goals. Although the Program Executive described the NEO Program's primary objective as finding as many asteroids as possible as quickly as possible, NASA has no structure or metric to evaluate the Program's effectiveness in reaching this objective.

Moreover, to meet the 2005 Authorization Act goal, the NEO Program provided support for ground-based telescopes with larger cameras and more sophisticated data acquisition, processing, and dissemination systems and obtained additional observation time at Pan-STARRS for detection of objects and at Arecibo for characterization. In addition, the Program funded research studies related to the mitigation of NEO hazards and supported demonstration projects such as ATLAS and the KaBOOM Project.

In addition to NEO-specific activities, the Program supports the work of other NASA initiatives, including the Asteroid Redirect Mission. NEO Program personnel also provide technical support for a Space Act Agreement with the B612 Foundation to help develop a privately funded, space-based, infrared telescope. In FY 2014, the Program also began funding the Antarctic Search for Meteorites Project (together with the NSF and the Smithsonian Institution) that sends researchers to Antarctica to collect and deliver meteorites to the Smithsonian Institution for permanent storage. Despite this increased activity, NASA has not changed or improved the NEO Program management structure and the Program has not established a plan to integrate the additional initiatives or demonstrate and track their contributions to attainment of NEO Program goals.

NASA Policy Directive (NPD) 7120.4D sets forth requirements for planning, tracking the progress of, and accomplishing mission objectives for Agency programs and projects.[24] In addition, NASA Procedural Requirements (NPR) 1080.1A requires research and technology program leads to demonstrate the relevance, quality, and performance of their program investments; demonstrate that their programs have well-conceived plans identifying goals, priorities, and linkages to national and stakeholder or beneficiary needs; and justify how funds will be allocated to ensure quality.[25] The NPR also requires programs to establish plans and processes to monitor and document performance, including appropriate outcome measures and milestones to track progress toward goals and assess whether funding needs to be increased or redirected.

Further, to achieve technical objectives within cost and schedule constraints, NPR 7120.8 requires project leads to establish a Work Breakdown Structure, a project schedule with milestones for each element in the Structure, and an allocation of the project's available resources necessary to achieve each milestone.[26] In addition, the milestones should be set at intervals sufficient to demonstrate steady progress toward achieving overall key decision points, as shown in Figure 6.

Because it lacks a program plan with associated cost estimates, NASA could not identify resources required to adequately support the Program, explain how some of the efforts to which the NEO Program contributes further Program goals, or predict when statutory goals would be met. For example, the NEO Program does not have an integrated plan for how it will contribute to the development or use NEOCam, the KaBOOM radar array, or information from the OSIRIS-Rex mission. Moreover, the manner in which the developmental paths for these projects are coordinated with NEO Program requirements is not clear since there is no integrated timeline or requirement for the projects to report milestones.

We believe the NEO Program would be better equipped to manage its various funding initiatives, support related NASA initiatives, and meet Program goals if it utilized clear, concise, verifiable, and valid requirements pertaining to NEO detection, characterization, and mitigation and followed NASA's policies for research programs and projects. Without an appropriate management and staffing structure, the Program Executive is unable to evaluate needs compared to requirements and effectively communicate those needs to stakeholders.

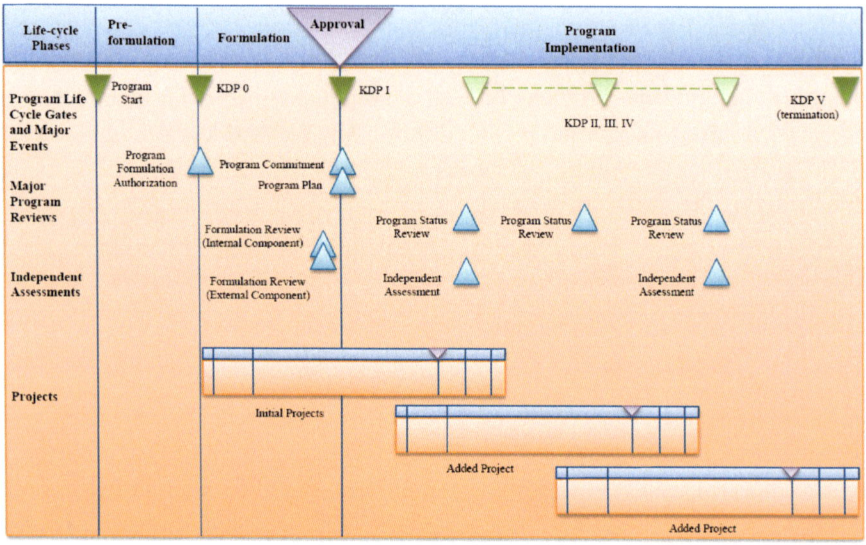

Note: Key decision point (KDP).
Source: NPR 7120.8.

Figure 6. Research and Technology Program Life Cycle.

NASA Needs to Improve Its Oversight of NEO Grants and Task Orders

NASA's controls for managing and overseeing costs associated with the NEO Program are inadequate to ensure proper accounting of NASA-funded grants and task orders. Specifically, due to a lack of personnel and internal controls the Program lacks adequate segregation of duties and sufficient oversight of grants and contracts by management. In addition, the Program used grants when contracts were the more appropriate procurement vehicle, thereby reducing NASA's ability to monitor grantees' performance and deliverables.

Inadequate Segregation of Duties

We found that contrary to effective internal control standards, the NEO Program Executive is responsible for or has significant input into all the primary elements of the award process. The Science Mission Directorate Management Handbook directs the Program Executive to review proposals for compliance with solicitations, identify and select panel members to review proposals, review panel recommendations, and make a recommendation for selection to the selection official – the Planetary Science Division Director.[27] According to the Program Executive, neither the Director nor the NASA Shared Services Center have ever deviated from his award recommendations. In addition to these responsibilities, the Program Executive also serves as the Contracting Officer's Technical Representative for all NEO Program awards. In that role, the Executive oversees, monitors, and evaluates the progress of all awards. Furthermore, the Executive reviews grantees' annual progress reports and approves requests for additional funding as necessary. Although we did not find any instances, this multi-responsibility role increases the risk of error and fraud in the NEO Program.

Government Accountability Office internal control standards provide that no single individual should control all important aspects of an event and key duties and responsibilities need to be divided among different individuals to reduce the risk of error or fraud in Federal programs.[28] This includes separating responsibility for authorizing, processing, recording, and reviewing transactions, as well as handling any related assets. Furthermore, an Association of Certified Fraud Examiners study reviewing 1,843 cases of occupational fraud in the United States and more than 100 other countries involving more than $18 billion in losses found that a lack of internal controls, including failure to segregate duties, contributed to fraud in 37.8 percent of cases.[29] Accordingly, having the Program Executive perform all of the duties outlined above increases the risk of error or fraud in the NEO Program.

Inadequate Oversight and Inappropriate Use of Funding Instruments

We reviewed all awards exceeding $900,000 funded by the NEO Program during FY 2013. These six grants, three task orders (contracts), and one interagency transfer accounted for 67 percent of the NEO budget that year (see Table 2).[30]

Table 2. Summary of Reviewed Grants, Task Orders, and Interagency Transfer

Task Title	Task Organization	Award Type	FY 2013 Funding
Near Earth Object Program Office	JPL	Task Order	$1,131,936
Primitive Body Navigation	JPL	Task Order	1,061,736

Counsel that stated, "...there is a basis to keep it a grant. In fact, it would be unnecessary and counterproductive to shift to a contract mechanism," and "... because NASA is supplementing support of a federal activity that is primarily funded and managed by NSF [National Science Foundation], which has chosen a grant mechanism."

However, in October 2011 the NSF did not use a grant, but awarded a 5-year, $42 million cooperative agreement to the Stanford Research Institute to manage, operate, and maintain the Arecibo Observatory. In our judgment, using contracts in these instances would have benefitted the Program and provided greater visibility into Arecibo's operations as well as provide assurance that the level of funding and performance was commensurate with contractual requirements. Moreover, given that funding for the NEO Program has grown ten-fold since 2009, we believe the Program needs more personnel to ensure proper oversight of associated research and procurement awards. The Program Executive agreed, suggesting the Program requires a minimum of four to six full-time equivalent positions to be fully effective.[34]

Lack of Formal Agreements with International and Federal Partners Hampers NASA's Ability to Accomplish Program Goals

Pursuant to Presidential Policy Directive, NASA is required to "[p]ursue capabilities, in cooperation with other departments, agencies, and commercial partners, to detect, track, catalog, and characterize near-Earth objects to reduce the risk of harm to humans from an unexpected impact...."[35] Although NASA has established two formal partnerships related to NEOs, lack of planning and resources has prevented the NEO Program from developing additional partnerships that could help achieve Program goals. Moreover, the NEO Program has not taken advantage of existing knowledge within NASA regarding forming partnerships. Expanded partnerships could increase the number of NEOs discovered, aid in development of mitigation strategies, and help ensure the burden of mitigating NEO threats does not fall disproportionately on NASA and the United States.

Existing Partnerships. Partnerships are an increasingly important component of NASA's overall mission strategy. With regard to NEOs, NASA has established formal partnerships with two organizations – the B612 Foundation and Planetary Resources. Through these partnerships, NASA supported development of a privately funded, space-based, infrared telescope by the B612 Foundation and an effort by Planetary Resources to create a crowdsourcing algorithm related to NEO survey work.

In addition to these formal partnerships, NASA has also entered into a number of informal international partnerships. For example, the NEO Program is a leading participant in the United Nations Committee on the Peaceful Uses of Outer Space. Over the past several years, a working group of the Committee's Scientific and Technical Subcommittee has studied NEOs and recently made recommendations to broaden and strengthen international cooperation and collaboration and called for national space agencies to form a group focused on designing a set of standard, agreed upon missions for deflecting NEOs.[36]

NEO Program initiatives have also resulted in informal partnerships. For example, the Catalina and Pan-STARRS observatories have taken first steps towards coordinating their survey efforts to ensure they are not duplicating efforts. In our judgment, the NEO Program Executive should replicate this model and establish a systematic process to improve observational efficiency and telescope utilization among international entities, intergovernmental organizations, and Program participants.

The NEO Program Could Leverage Agency Experience to Form Partnerships. A number of NASA programs and offices, including the International Space Station Program, NASA's Office of International and Interagency Relations, and the Office of the Chief Technologist, have formal agreements or help promote partnerships with international and domestic partners regarding various aspects of their goals and respective missions. The NEO Program may be able to leverage the knowledge of these organizations to establish agreements of its own. For example, NASA established the Global Exploration Roadmap with foreign space agency members of the International Space Exploration Coordination Group.[37] One of the Global Exploration Roadmap's goals is "Enhancing Earth Safety" by contributing to the "collaborative pursuit of planetary defense and orbital debris management mechanisms." Citing the NEO Program as one of NASA's efforts in fulfilling this goal could create opportunities for partnerships between member countries and maximize access to and development of technology as well as augment mitigation investments.

Partnerships Could Expand Access to Technological Assets and Spread the Burden of Detection and Mitigation. The 2005 Authorization Act directed NASA to detect NEOs and develop mitigation strategies to determine how to alter the path of or destroy PHOs. However, NASA's lack of NEO Program partnerships and insufficient staff to pursue such partnerships limits the Nation's ability to discover NEOs and weakens the Agency's ability to plan for effective mitigation. For example, establishing formal partnerships with the Department of Defense and the NSF could give the NEO Program access to additional Earth-based telescopes, thereby enhancing sky-viewing capabilities and the ability to detect, track, and characterize a greater number of NEOs and PHOs and increasing the Program's ability to plan for and implement effective mitigation strategies. Although, the NEO Program Executive told us he has had informal discussions with these agencies regarding use of their telescopes, we believe the Program Executive should leverage NASA's Office of International and Interagency Relations and the Office of Chief Technologist to expedite formation of agreements.

The NEO Program lacks a plan for establishing agreements with domestic and foreign entities to leverage capabilities or mitigate NEO threats and has not developed milestones to measure progress toward established goals. Due to the variety of possible sizes, trajectories, and warning times for PHOs, developing an appropriate mitigation response is extremely challenging. For example, PHOs larger than a few hundred meters in diameter would require enormous energies to deflect or fragment, as well as cooperation on the part of multiple nations. Although NEO Program managers stated they would alert the Federal Emergency Management Agency in the event of an impending PHO threat to the United States or the Department of State if predicted impact was outside the United States, there is no interagency agreement or formal procedure between the NEO Program and either agency, or with the Department of Defense and Department of Energy in the event a nuclear option is required.

NEO Program management told us that if given the opportunity, they would like to form partnerships with the Russian and the European space agencies. The Russian Space Agency is considering forming a separate agency to monitor NEOs and the European Space Agency is building a 1-meter telescope intended to cover the entire sky every few nights to search for NEOs. Other countries have similar initiatives in planning or development.

Recommendations, Management's Response, and Evaluation of Management's Response

To improve NASA's efforts to discover, characterize, and mitigate NEO threats, we recommended the Associate Administrator for the Science Mission Directorate:

Recommendation 1. Develop a formal NEO Program in accordance with NASA policies that includes a strategic plan to obtain statutory and programmatic goals, integrated master schedule, and cost estimates.

Management's Response. The Associate Administrator concurred and agreed that managing the Program in accordance with NASA policies will maximize results and may reduce costs. The Associate Administrator expects the new Program plan will be in place by September 1, 2015.

Evaluation of Management's Response. Management's comments are responsive; therefore, the recommendation is resolved and will be closed upon verification and completion of the proposed corrective actions.

In addition, we recommended the Associate Administrator direct the NEO Program Executive to:

Recommendation 2. Develop and implement requirements, procedures, and internal controls to address program deficiencies.

Management's Response. The Associate Administrator concurred and expects to develop requirements, procedures, and internal controls consistent with Agency policy by March 1, 2015.

Evaluation of Management's Response. Management's comments are responsive; therefore, the recommendation is resolved and will be closed upon verification and completion of the proposed corrective actions.

Recommendation 3. Perform a full-time equivalent analysis to determine the appropriate number of staff needed to manage, oversee, and administer the Program and associated funding instruments.

Management's Response. The Associate Administrator concurred, stating that a full-time equivalent analysis will be conducted by March 1, 2015.

Evaluation of Management's Response. Management's comments are responsive; therefore, the recommendation is resolved and will be closed upon verification and completion of the proposed corrective actions.

Recommendation 4. Develop a plan to establish formal, documented partnerships with domestic and international agencies to leverage resources and complementary technologies.

Management's Response. The Associate Administrator concurred, stating that the Program is on the proper course to establish the appropriate domestic and international partnerships and that a plan to establish formal, documented partnerships with domestic and international agencies will be developed by September 1, 2015.

Evaluation of Management's Response. Management's comments are responsive; therefore, the recommendation is resolved and will be closed upon verification and completion of the proposed corrective actions.

Recommendation 5. Establish a systematic oversight process pursuant to which NASA-funded observatories are required to coordinate with each other in order to avoid duplication of efforts.

Management's Response. The A ssoci ate A dmi ni strator concurred, stating the Program has al ready taken steps to ensure project activities are coordinated and will continue to do so. He also stated that a strategy for coordination of activities will be documented in the N EO Program Plan by September 1, 2015.

Evaluation of Management's Response. Management's comments are responsive; therefore, the recommendation is resolved and will be closed upon verification and completion of the proposed corrective actions.

APPENDIX A

Scope and Methodology

We performed this audit from September 2013 through August 2014 in accordance with generally accepted government auditing standards. Those standards require that we plan and perform the audit to obtain sufficient, appropriate evidence to provide a reasonable basis for our findings and conclusions based on our audit objectives. We believe the evidence obtained provides a reasonable basis for our findings and conclusions based on our audit objectives.

We reviewed research and technology program requirements, grant and cooperative agreement requirements contained in the following NASA policies, as well as Center-specific policies that corresponded with NASA guidance, and public laws:

- National Aeronautics and Space Act, Pub. L No. 111–314, December 18, 2010.

- NPD 1080.1B, "Policy for the Conduct of NASA Research and Technology (R&T)," July 9, 2008.
- NPD 7120.4D, "NASA Engineering and Program/Project Management Policy," March 16, 2010.
- NPR 7120.8, "NASA Research and Technology Program and Project Management Requirements, (w/change 3 dated 04/18/13)," February 5, 2008.
- NPR 1080.1A, "Requirements for the Conduct of NASA Research and Technology (R&T)," May 30, 2008.
- NPR 9680.1A, "NASA Financial Management of Grant and Cooperative Agreements," November 10, 2011.
- Headquarters Science Mission Directorate (SMD) Handbook, February 8, 2008.
- Pub. L. No. 109–155, Subtitle C – George E. Brown, Jr., Near Earth Objects Survey Act, December 30, 2005.
- House of Representatives 6063, National Aeronautics and Space Administration Authorization Act of 2008 (Enrolled as Agreed to or Passed by Both House and Senate), Title VIII – Near Earth Objects, Section 801, "Reaffirmation of Policy."
- 31 U.S.C. § 6303, "Using Procurement Contracts."
- 31 U.S.C. § 6304, "Using Grant Agreements."
- 14 C.F.R. § 1260.13, Section A, "Grants and Cooperative Agreements."

We reviewed internal control requirements contained in the Government Accountability Office's "Standards for Internal Controls in the Federal Government," November 1999, and Office of Management and Budget Circular A-123, "Management Responsibility for Internal Controls," December 21, 2004.

To determine whether the NEO Program devised an effective strategy to meet the findings and recommendations of the NRC, we interviewed NASA personnel, scientists, and astrophysicists about their roles and responsibilities. We assessed the NEO Program oversight procedures to determine whether oversight reviews were conducted as required. We also conducted fieldwork at the Minor Planet Center in Cambridge, Massachusetts, the Catalina Sky Survey in Arizona, and the KaBOOM Project at Kennedy Space Center. Further, we attended two conferences:

- Small Bodies Assessment Group, Washington, DC, January 2014
- International Asteroid Warning Network hosted by the Minor Planet Center, January 2014

To assess whether funding and resources were adequately allocated, we judgmentally selected for review 10 funding instruments that exceeded $900,000, which included 6 grants, 3 task orders, and 1 interagency transfer.

To determine whether the NEO Program's monitoring systems were adequately maintained and updated and that the data was effectively processed, communicated, and tracked, we reviewed the Minor Planet Center and JPL systems and databases.

To assess NEO Program mitigation and partnership strategies, we interviewed scientists receiving NEO grants to determine whether any plans for mitigation existed and any agreements executed to form partnerships within the Federal Government or with international partners.

Use of Computer-Processed Data. We used data from the Research and Analysis Program Tracking of Resources system to perform this audit.[38] Although we did not test the general or application controls of the system, we did compare the information in the key data fields with our sample of funding instruments and supporting documents for the data and determined that the data was valid and reliable to support our objectives and conclusions.

Review of Internal Controls

We performed an assessment of the internal controls associated with NASA's NEO Programs. Throughout the audit we reviewed controls associated with the audit objectives and determined that NASA's internal controls and program oversight were inadequate to ensure proper segregation of duties and oversight and monitoring of NEO funding instruments.

Prior Coverage

During the past 5 years, the NASA Office of Inspector General (OIG) and the Government Accountability Office have not issued any reports of particular relevance to the subject of this report. However, we issued NASA OIG, "NASA's Grant Administration and Management" (IG-11-026, September 12, 2011), which identified systemic deficiencies in the administration and management of NASA's grant program. We also relied on NASA OIG, "Federal Information Security Management Act: Fiscal Year 2013 Evaluation" (IG-14-004, November 13, 2013), to review systems and databases at JPL. Unrestricted reports can be accessed over the Internet at http://oig.nasa.gov/audits/reports/FY14.

APPENDIX B

Neo Program Assets and Related Initiatives

The NEO Program operations consist of detection and notification, characterization and cataloging, and mitigation.

Detection and Notification. The NEO Program provides funds to three survey teams that operate five ground-based telescopes to detect NEOs. The majority (73 percent) of the observations reported to the Minor Planet Center come from three survey teams:

- The *Catalina Sky Survey (Catalina)*, including the Mount Lemmon Survey and the Siding Spring Survey, is based at the University of Arizona. A successful discovery program due to its comprehensive sky coverage and on site follow-up observation capabilities, in 2013, Catalina discovered more than 600 NEOs. Catalina observation time is split between detection and follow-up with approximately 20 percent devoted to post-discovery, follow-up observations that allow computation of an object's orbit.
- The *Lincoln Near-Earth Asteroid Research (LINEAR)* program at White Sands, New Mexico, is run by the Massachusetts Institute of Technology and leads in the discovery of NEOs larger than 1 kilometer. LINEAR uses electro-optical sensor technology developed for U.S. Air Force space surveillance applications to discover NEOs and is largely responsible for NASA successfully meeting the goal of finding 90 percent of the 1-kilometer and larger NEOs. LINEAR is fully dedicated to the NEO Program and schedules 20 to 25 nights per month for observations.
- The *Panoramic Survey Telescope and Rapid Response System 1 and 2 (Pan-STARRS)* discovery telescope, run by the University of Hawaii's Institute for Astronomy, is a 1.8-meter telescope with a wide field of

view and incorporating a very large camera on Haleakala in Maui, Hawaii (see Figure 7). The NEO Program funds approximately 11 percent of its total available observation time.

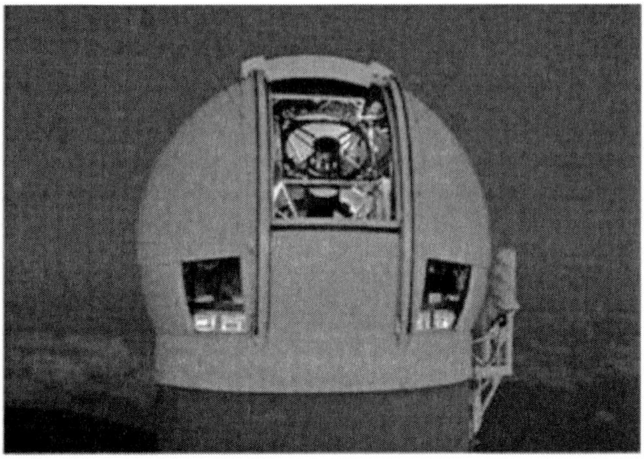

Source: Institute for Astronomy, University of Hawaii. Photo by Rob Ratkowski.

Figure 7. Pan-STARRS.

To supplement the three Earth-based surveys, in September 2013, NASA reactivated and reprogrammed the Wide-field Infrared Survey Explorer (WISE) telescope to search for NEOs.[39] The telescope – renamed NEOWISE – is a space-based infrared telescope that can detect the ambient heat from asteroids. It is particularly useful for discovering dark colored asteroids that are extremely difficult to detect with ground-based conventional telescopes.

Characterization and Cataloging. The interaction of radar signals with the surface of a NEO yields information about its physical characteristics, such as the NEO's size, shape, rotation characteristics, and whether it has any moons. For example, radar observations can be used to estimate the roughness of the top layer of a NEO's surface, which could distinguish between stony and metallic compositions and may be used to estimate the porosity of a NEO's surface and help in understanding its composition for developing mitigation strategies.[40]

Source: NASA.

Figure 8. Goldstone 70-meter Telescope.

The NEO Program provides funding to two radar systems:

- *Goldstone Solar System Radar (Goldstone)*, located in the Mojave Desert in Southern California, is part of NASA's Deep Space Network and is operated by JPL under contract with NASA (see Figure 8). Comprised of a fully steerable 70-meter-diameter antenna, this radar has a significant capability for observing NEOs in 80 percent of the sky north of -35° latitude over the course of a day. The Goldstone antenna's primary mission is spacecraft communications and therefore is available for astronomy observations only a few percent of its time.
- *The Arecibo Observatory (Arecibo)* in Puerto Rico is part of the National Astronomy and Ionosphere Center, a national research center operated by the Stanford Research Institute under a cooperative agreement with NSF. NASA provides additional support through grants with Universities Space Research Association. Its chief feature is a fixed 305-meter-diameter spherical antenna that allows coverage directly overhead. Observing time at Arecibo is awarded on a competitive basis from proposals that are normally submitted quarterly. Arecibo is also available for "urgent" target-of-opportunity observations on short notice and, in a small number of instances, has

been used for radar observations of NEOs within 24 hours of their discovery.

Mitigation. The NEO Program allocated about $1 million per year to study mitigation strategies, such as:

- NEO Observations Program Mitigation Effects grant, which is a comprehensive study of impact and explosion mitigation technologies for threatening asteroids;
- NASA Innovative Advanced Concepts Program, which studies entirely new or radically different mitigation strategies from American innovators and entrepreneurs; and
- Kinetic Impactor Demonstration Mission Studies, which include a mission concept for the use of two independent spacecraft, one impactor and one rendezvous probe, where the rendezvous probe would observe the impact of an asteroid and measure any change in the asteroids relative orbit.

Other Related NASA and Federal Government NEO Initiatives

The KaBOOM Project is a capability demonstration project at Kennedy Space Center (see Figure 9). KaBOOM technology would be complimentary to other radar technologies by providing tracking and characterization at Ka-band frequencies for NEOs at much further distances and far higher resolution than currently available. Compared to NASA's Goldstone 70-meter antenna, which can track an object that is about 9 million miles away (in X-band frequencies) using an array of multiple radar dishes, KaBOOM may be able to track objects 47 million miles or more away – potentially providing more time for a mitigation strategy to be developed in the event of a PHO threat. However, Project managers stated development of the technology and system would take approximately 7–8 years before it would be ready to deploy in a 24/7 operational radar array.

The Asteroid Terrestrial-impact Last Alert System (ATLAS) is being developed by the University of Hawaii and intended to be operational in 2016. The array of eight telescopes is designed to offer a 1-week warning for a 50-yard-diameter asteroid, or "city killer," and 3 weeks for a 150-yard-diameter "county killer," which provides sufficient time to evacuate the area, take measures to protect buildings and other infrastructure, and be alert to a tsunami danger generated by ocean impacts.

The NEO Camera (NEOCam) is a wide-field camera operating at thermal infrared wavelengths being developed at JPL and intended to be integrated onto a space-based telescope for the purpose of improving NEO detection capabilities. NEOcam is intended to be launched to a position about four times the distance between Earth and the moon, where it could observe NEOs without the impediments of cloud cover or daylight. However, there is presently no funding available to design and manufacture a spacecraft to carry the NEOCam payload.

Source: NASA.

Figure 9. KaBOOM Radar Array.

The Asteroid Redirect Mission that is being developed by NASA's Human Exploration and Operations Mission Directorate aims to fulfill the President's call to send astronauts to a near-Earth asteroid by 2025. The mission objective is to identify, capture, redirect, and sample an asteroid with a mass of up to about 1,000 tons into a stable lunar orbit by the first half of the next decade. NASA has stated that the mission would provide invaluable new data on the threats asteroids pose to Earth and how they could be mitigated.

The Origins-Spectral Interpretation-Resource Identification-Security-Regolith Explorer (OSIRIS-Rex) mission, planned for launch in 2016, is designed to rendezvous with an asteroid and use a robotic arm to collect samples that could better explain our solar system's formation and how life began. The mission will be the first U.S. mission to carry samples from an asteroid back to Earth. Because the spacecraft will spend a significant amount

In the report, the Office of Inspector General (OIG) makes five recommendations intended to improve NASA's efforts to discover, characterize, and mitigate near-Earth orbit (NEO) threats. NASA's response to the OIG's recommendations outlined in the report, including planned corrective actions, follows.

To improve NASA's efforts to discover, characterize, and mitigate NEO threats, the OIG recommends that the Associate Administrator for the Science Mission Directorate (SMD):

Recommendation 1: Develop a formal NEO Program in accordance with NASA policies that includes a strategic plan to obtain statutory and programmatic goals, integrated master schedule, and cost estimates.

Management's Response: SMD concurs with this recommendation. The assets currently used, as well as additional capabilities needed, to detect, track, and characterize the hazardous NEO population can best be described as a loosely coupled system of systems that are managed, designed, and operated by a variety of science institutions and personnel – almost all within a scientific research background and setting. These elements operate at or very near the forefront of our scientific understanding of these smallest constituents of our solar system and the engineering of the systems is likewise at the edge. We agree that the proposed approach to manage this effort, in accordance with NPR 7120.8, NASA Research and Technology Program and Project Management Requirements, will maximize the results and offer the potential to reduce contractual costs while providing the flexibility to maintain individually customized research approaches under over-arching strategies established at a program level.

Estimated Completion Date: The new program will be established by September 1, 2015.

The OIG also recommends that the Associate Administrator for SMD should direct the NEO Program Executive to:

Recommendation 2: Develop and implement requirements, procedures, and internal controls to address program deficiencies.

Management's Response: SMD concurs with establishing a program that follows the Agency's requirements, procedures, and internal controls in exercising the Agency's fiduciary responsibilities and firmly believes that the NEO Observations projects that are currently funded already follow NASA's standard procedures for awarding research grants and contracts, with multiple levels of review and independent internal controls in the award and funding of our research activities.

Estimated Completion Date: Requirements, procedures, and internal controls for the new program will be identified by March 1, 2015.

Recommendation 3: Perform a full-time equivalent (FTE) analysis to determine the appropriate number of staff needed to manage, oversee, and administer the Program and associated funding instruments.

Management's Response: SMD concurs with this recommendation.

Estimated Completion Date: An FTE analysis will be conducted by March 1, 2015 and incorporated into plans for the program.

Recommendation 4: Develop a plan to establish formal, documented partnerships with domestic and international agencies to leverage resources and complementary technologies.

Management's Response: SMD concurs with this recommendation, and it will be incorporated into the NEO Program Plan. It is our view that the program is on the proper course to establish both the appropriate domestic and international partnerships. While NEO activities are very new to our potential partners, there have been considerable efforts within U.S. Government interagency working groups, and international coordination efforts within the constructs of the United Nations Committee On Peaceful Uses of Outer Space Working Group on NEOs are highly documented, all of which are worked in coordination with the NASA Office of Interagency and International Relations.

Estimated Completion Date: This plan will be developed by September 1, 2015.

> **Recommendation 5:** Establish a systematic oversight process pursuant to which NASA-funded observatories are required to coordinate with each other in order to avoid duplication of effort.
>
> **Management's Response:** SMD concurs with the intent of this recommendation. As noted in the audit report, SMD has already taken steps to ensure project activities are coordinated and will continue to do so in the future. Strategy for coordination of activities will be documented in the program plan. This is not the same as coordinating nightly observations of survey teams, where we believe the existing competitive environment across all of our assets is maximizing the identification of new NEOs through continuous improvement for increased sky coverage, improved detection limits, moving object algorithms and astrometric accuracy.
>
> **Estimated Completion Date:** This oversight process will be in place by September 1, 2015, when the new program is established.

At the request of the OIG, we have reviewed the draft report for information that we believe should not be publicly released and have not identified any concerns in this regard.

Again, thank you for the opportunity to review and comment on the subject draft report. If you have any questions or require additional information regarding this response, please contact Peter Meister at (202) 358-1557.

Dr. John M. Grunsfeld

Appendix E

Report Distribution

National Aeronautics and Space Administration

Administrator
Associate Administrator
Chief of Staff
Associate Administrator for the Science Mission Directorate
Director, Jet Propulsion Laboratory

Non-NASA Organizations and Individuals

Office of Management and Budget
Deputy Associate Director, Energy and Science Division
Branch Chief, Science and Space Programs Branch
Government Accountability Office
Director, Office of Acquisition and Sourcing Management

Congressional Committees and Subcommittees, Chairman and Ranking Member

Senate Committee on Appropriations
Subcommittee on Commerce, Justice, Science, and Related Agencies
Senate Committee on Commerce, Science, and Transportation
Subcommittee on Science and Space
Senate Committee on Homeland Security and Governmental Affairs
House Committee on Appropriations
Subcommittee on Commerce, Justice, Science, and Related Agencies
House Committee on Oversight and Government Reform
Subcommittee on Government Operations
House Committee on Science, Space, and Technology
Subcommittee on Oversight
Subcommittee on Space

Final report released by:
Paul K. Martin Inspector General
Major Contributors to the Report:
Raymond Tolomeo, Director,
Science and Aeronautics Research Directorate
Ronald Yarbrough, Project Manager

Jiang Yun Lu, Auditor
Anh Doan, Management Analyst
Noreen Khan-Mayberry, Ph.D., Technical Evaluator
Sarah McGrath, Report Process Manager

End Notes

[1] The Spaceguard Survey, "Report of the NASA International Near-Earth-Object Detection Workshop," January 25, 1992.
[2] United States Congress, H. R. 4489, Report No. 103–654, August 3, 1994.
[3] In various documents, NASA refers to this research effort as the Near-Earth Object Observations Program. For ease of reference, we use "NEO Program" or "the Program" in this report.
[4] Pub. L. No. 109–155, December 30, 2005 (codified at 42 U.S.C. § 16691).
[5] The Asteroid Redirect Mission, developed by NASA's Human Exploration and Operations Mission Directorate, aims to fulfill the President's call to send astronauts to a near-Earth asteroid. The mission objective is to identify, capture, redirect into a stable lunar orbit, and sample an asteroid with a mass up to 1,000 tons by the first half of the next decade.
[6] NASA Procedural Requirements (NPR) 1080.1A, "Requirements for the Conduct of NASA Research and Technology (R&T)," May 30, 2008. NPR 7120.8, "NASA Research and Technology Program and Project Management Requirements (w/change 3 dated 04/18/13)," February 5, 2008.
[7] Government Accountability Office, "Standards for Internal Control in the Federal Government" (AIMD-00-21.3.1, November 1999).
[8] 31 U.S.C. § 6303, "Using Procurement Contracts."
[9] Astronomers Carl Gustav Witt and Auguste Charlois are credited with discovering the first NEO in 1898.
[10] The Chelyabinsk meteor was the largest reported since 1908 when a meteor hit Tunguska, Siberia, and flattened more than 772 square miles of forest. Scientists believe that between 9,000 and 13,000 pounds of the Chelyabinsk meteor fell to Earth in various size fragments. One fragment weighing approximately 1,400 pounds was recovered from a lake about 45 miles east of the city.
[11] As discussed in more detail later in this report, characterizing the physical and orbital properties of a NEO is required to determine its threat potential and to plan any necessary mitigation efforts.
[12] The Spaceguard Survey, "Report of the NASA International Near-Earth-Object Detection Workshop," January 25, 1992.
[13] United States Congress, H. R. 4489, Report No. 103–654, August 3, 1994.
[14] Pub. L. No. 109–155, December 30, 2005 (codified at 42 U.S.C. § 16691).
[15] 42 U.S.C. § 2451(g).
[16] NRC, "Defending Planet Earth: Near Earth Object Surveys and Hazard Mitigation Strategies," 2010.
[17] NASA refers to this research effort in various documents as the Near-Earth Object Observations Program. For ease of reference, we use "NEO Program" or "the Program" in this report.

[18] The JPL NEO Office works with the European Space Agency's Near Earth Objects - Dynamic Site on issues of mutual interest.

[19] In FY 2009, the NEO Program received an additional $1.9 million from the Science Mission Directorate's Planetary Research and Analysis budget.

[20] Launched in December 2009, NASA placed WISE into hibernation in February 2011 after it had completed its original mission of surveying the sky in infrared light to look for dwarf stars and undiscovered asteroids.

[21] The "slow-push" and "slow-pull" methods involve a propulsion device or other large body positioned near an object to change its orbit and avoid collision with Earth. The technique takes considerable time to implement – on the order of decades – and would only be effective for objects with diameters no larger than 100 meters. Kinetic impactors change a NEO's orbit by striking it directly at high speed. The nuclear option is the last resort because of concerns over uncontrolled fragmentation.

[22] The KaBOOM Project is an array of radar dishes that could provide tracking and characterization of NEOs at much further distances and far higher resolution than currently available. ATLAS is a NEO detection system comprised of eight ground-based telescopes. NEOCam uses heat to detect NEOs. See Appendix B for more detail on these systems.

[23] The objective of the Asteroid Redirect Mission is to identify, capture, and redirect by the first half of the next decade a small asteroid into a stable lunar orbit for further exploration. In 2016, NASA intends to launch OSIRIS-Rex to an asteroid and use a robotic arm to collect samples that could better explain our solar system's formation and how life began. Because the spacecraft will spend a significant amount of time at the asteroid – up to 2.5 years – the mission could help improve asteroid orbit predictions.

[24] NPD 7120.4D, "NASA Engineering and Program/Project Management Policy," March 16, 2010.

[25] NPR 1080.1A, "Requirements for the Conduct of NASA Research and Technology (R&T)," May 30, 2008.

[26] NPR 7120.8, "NASA Research and Technology Program and Project Management Requirements (w/change 3 dated 04/18/13)," February 5, 2008.

[27] Final awards are typically made by the NASA Shared Services Center – a partnership between NASA and a contractor that consolidates certain support functions such as financial management, human resources, information technology, and procurement.

[28] Government Accountability Office, "Standards for Internal Control in the Federal Government" (AIMD-00-21.3.1, November 1999).

[29] Association of Certified Fraud Examiners, "Report to the Nations on Occupational Fraud and Abuse: 2010 Global Fraud Study" (2010).

[30] Interagency transfer is the transfer of funds from one Federal agency to another for the purpose of conducting research.

[31] The SST is primarily intended to detect and track small objects in geosynchronous orbit, approximately 22,000 miles above Earth.

[32] 31 U.S.C. § 6304, "Using Grant Agreements."

[33] 31 U.S.C. § 6303, "Using Procurement Contracts."

[34] Full-time equivalent refers to the number of hours worked that add up to one full-time employee working an 8-hour day, 5 days a week for 52 weeks.

[35] Presidential Policy Directive 4, "National Space Policy of the United States of America," June 28, 2010.

[36] United Nations Committee on the Peaceful Uses of Outer Space, *Report of the Scientific and Technical Subcommittee on its fifty-first session, held in Vienna from 10 to 21 February*

2014 (Vienna: United Nations, 2014), available from http://www.oosa.unvienna.org/pdf/reports/ac105/AC105 1065E.pdf (last accessed July 28, 2014).

[37] The International Space Exploration Coordination Group is comprised of 14 space agencies with the goal to advance a long-range human space exploration strategy.

[38] The Research and Analysis Program Tracking of Resources system is a Science Mission Directorate web-based tool for tracking approved investigations and specifying the award funding vehicles and amounts.

[39] WISE launched in December 2009 and was placed in hibernation in February 2011 after completing its mission of surveying the sky in infrared light and discovering dwarf stars and new asteroids.

[40] Porosity is the ratio of the volume of space to the total volume of a rock.

- What are the most viable efforts to focus on in the next 5 to 7 years that will yield the most progress in identifying and cataloging the NEO threat?
- Are we tracking the right size objects, specifically the ones that can cause significant harm? Once we identify an object, what are our means of tracking it?
- What are our contingencies and mitigation capabilities if we determine there is a threat from a NEO impact?

The Science, Space, and Technology Committee has been on the forefront of the issues surrounding Near Earth Objects. For example, the NASA Authorization Acts of 2000 and 2005 directed NASA to conduct a survey of the population of NEOs and study mitigation plans. NASA estimates 20,000 potentially hazardous asteroids orbit within the vicinity of the Earth.

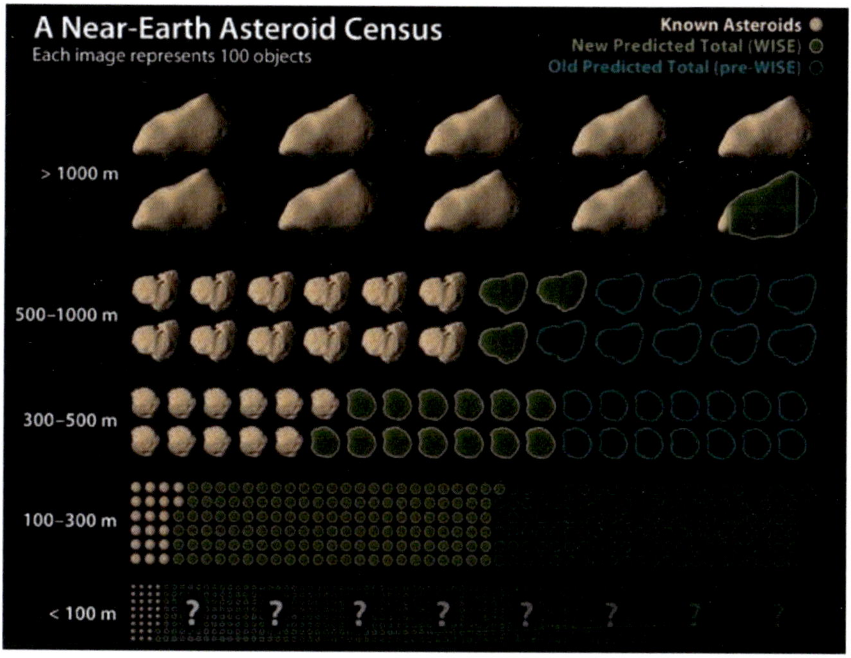

NASA NEO Asteroid Size Model

Credit: NASA/JPL-Caltech

This chart illustrates how infrared is used to more accurately determine an asteroid's size.

Department of State's diplomatic mechanisms would come into play for international communications as needed.

With regard to risk mitigation, the Administration is committed to exploring and developing the capabilities and techniques necessary to protect the Earth in general, and the United States in particular, from NEO threats and implementing a collision mitigation campaign if necessary and appropriate. In 2008, the Executive Office of the President collaborated with NASA and the U.S. Air Force to run the first-ever disaster and deflection exercise, which included members from the National Security Staff, Joint Chiefs of Staff, Office of the Under Secretary of Defense for Policy, Missile Defense Agency, the Defense Threat Reduction Agency, the Department of Defense (DoD) National Security Space Office (now the Executive Agent for Space Staff), and the Department of Energy. The National Research Council and the NASA Advisory Council have also provided helpful recommendations and guidance on research priorities in impact mitigation techniques.

Among the highlighted needs are an improved understanding of NEO characteristics to enable more refined impact experiments; enhanced computer simulations; crewed and uncrewed *in situ* asteroid investigations; and further research and capabilities development in the domain of deflection, including explosive technologies, and impact scenarios (including design reference missions and gaming exercises). DoD and NASA have already shown tremendous leadership by taking the initiative to run multi-agency disaster and deflection exercises, and by collaborating in the development of an international disaster and deflection response scenario for the upcoming Planetary Defense Conference hosted by the International Academy of Astronautics in Flagstaff, Arizona.

In summary, the Administration, with the support of Congress, has taken many positive steps to improve NEO detection capabilities, including meeting the 90 percent detection goal for one-kilometer asteroids. Much more needs to be done, however, and it is important to note that a challenge on the scale of planetary defense cannot be met by any single nation or government alone. Rather it will be critical going forward that the Federal Government cooperate closely with domestic partners in industry, academia, and other sectors as well as with foreign governments and international organizations to achieve our shared goal of scientific discovery, exploration, and risk mitigation.

I thank the Committee for its continued support and interest in this issue and I look forward to continuing to work with you on it. I will be pleased to take any questions Members may have.

In: Near-Earth Objects
Editor: Leon Sinclair
ISBN: 978-1-63484-067-5

Chapter 3

Statement of Dr. John P. Holdren, Director, Office of Science and Technology Policy, Executive Office of the President of the United States. Hearing on "Threats from Space: A Review of U.S. Government Efforts to Track and Mitigate Asteroids and Meteors, Part 1"*

Chairman Smith, Ranking Member Johnson, and Members of the Committee, I am pleased to be here with you today to discuss the status and coordination of U.S. activities to detect, track, and characterize near-Earth objects (NEO) and to develop the capability to deflect any of dangerous size discovered to be on a collision course with Earth. This is a very timely topic — as underscored by the asteroid explosion over Russia on February 15 and the close flyby of an even larger asteroid the same day — and I am looking forward to sharing with you the Administration's perspective on this issue.

I want to start by acknowledging the emphasis that the Congress has placed on understanding and mitigating the threat from NEOs. I thank you for working

* This is an edited, reformatted and augmented version of a statement presented March 19, 2013 before the House Committee on Science, Space, and Technology.

hundred kilotons of TNT equivalent needed to do significant damage at the surface. (The asteroid that exploded over Chelyabinsk on February 15 is estimated to have been about 17 meters in size.)

The 1908 explosion over Tunguska in Siberia, which leveled trees over an area of more than 2,200 square kilometers (850 square miles), is thought to have been caused by a stony asteroid between 45 and 60 meters in diameter, imparting between 10 and 20 megatons of TNT equivalent. Such a strike is believed to be a once-in-a-thousand-year event (or, put differently, having a 1 in 1,000 chance of occurring in any given year). An energy release of this size could cause hundreds of thousands of casualties and massive destruction if it occurred over an urban area. The probability of *this* occurring, however, is much smaller than the one-in-one-thousand odds I just mentioned, inasmuch as land covers only 30 percent of the area of the planet and cities only 2-3 percent of the land area. Of course, a similarly sized or even larger asteroid that made it to the surface intact could cause significant damage even if it hit the ocean, by virtue of the resulting tsunami.

Depending on its composition and velocity, an asteroid of 140 meters in diameter could have an impact energy in the range of 50 to 500 megatons of TNT equivalent and would be capable of causing destruction over a large region. The probability of a strike of this magnitude has been estimated at about 1 in 30,000 per year. As I noted earlier, it is believed that more than 90 percent of all NEOs 1 kilometer or greater in diameter have now been identified. None of those identified so far appears to pose a risk of collision, which is fortunate as the impact of an object of this size would release between 20,000 and 200,000 megatons of TNT equivalent and likely imperil all of civilization. The object that impacted the Earth just off the Yucatan Peninsula 65 million years ago, and that is believed to have been responsible for mass extinctions across the planet, was an asteroid estimated to have been 10 kilometers in diameter.

Recent Examples

In recent years, several NEOs have made close passes by Earth. Two unrelated asteroids, estimated to have been between six and 20 meters in diameter, passed between the Earth and the Moon on September 8th, 2010. Neither posed a risk of striking the Earth, but they served as a reminder that these kinds of close flybys are not rare. It's estimated that almost every day, at

least one 10-meter near-Earth asteroid (part of the undiscovered p about 50 million) passes the Earth inside the orbit of the Moon.

The Chelyabinsk asteroid strike of February 15 lit up the sky a a series of shock waves that shattered windows over a Approximately 1,500 people were injured and 60 were hosp regional governor said that two-thirds of the injuries were light w shattered glass and other materials. More than 4,000 buildings we According to NASA estimates, the velocity of approach of th asteroid was about 18 kilometers per second and its energy was kilotons of TNT equivalent. These estimates enable us to corresponding mass of 11,000 metric tons. Early measurements i following the event underestimated the energy (and therefore the n asteroid, but with the addition of data in the following days bot ground and from orbiting satellites that witnessed the asteroid's energy measurement was refined. The entire event from the atmospheric entry to its disintegration at an altitude of about 20 took only 32.5 seconds. Thus far, approximately 50 small meteorite from this explosion have been found. One piece blasted a six-mete hole in the ice covering one of Siberia's many lakes.

On the same day as the unexpected Chelyabinsk event, the asteroid 2012 DA14 safely passed nearly 27,700 kilometers (17,2 from Earth, a close flyby that had been predicted many months in adv asteroid had been tracked since its February 2012 discovery. Tracking orbit calculations made by the various Federal agencies with responsi this area made it clear over a year ago that this asteroid did not pose a Earth, the International Space Station, or satellites in orbit. This even researchers to measure the object's path and orbit with greater p improving estimates of future near-Earth passes.

Analysis of data collected during these two contemporaneou indicates that the asteroid that exploded over Chelyabinsk was almost unrelated to the larger asteroid 2012 DA14; the smaller asteroid's tr was not consistent with its being a fragment that came off the larger notable that while the Chelyabinsk "fireball" was unusual for its s visibility, thousands of smaller strikes that still explode with enough e produce a fireball (defined as being brighter than the brightest planet each day. Most fireballs are not noticed because they occur over oc uninhabited regions or are masked by daylight. Nearly all of them are by objects too small to be detected before they enter Earth's atmosphere also too small to do damage on Earth's surface.

End Notes

[1] Only the largest asteroids — those few with diameters in the range of hundreds of kilometers — are spherical. Smaller ones are of irregular shape.

[2] Most asteroids with orbits crossing that of the Earth would impact with velocities between 12 and 20 kilometers per second, hence the choice of 16 kilometers per second as an intermediate figure.

In: Near-Earth Objects
Editor: Leon Sinclair

ISBN: 978-1-63484-067-5

Chapter 4

STATEMENT OF GENERAL WILLIAM L. SHELTON, COMMANDER, AIR FORCE SPACE COMMAND. HEARING ON "THREATS FROM SPACE: A REVIEW OF U.S. GOVERNMENT EFFORTS TO TRACK AND MITIGATE ASTEROIDS AND METEORS, PART 1"*

Chairman Smith, Ranking Member Johnson, and Members of the Committee, thank you for the opportunity to appear before you today to discuss Air Force Space Command's role in monitoring activity in the space domain. Space situational awareness underpins the entire spectrum of space activities and Air Force Space Command's focus is on providing forces and capabilities to United States Strategic Command (USSTRATCOM) to detect, track, identify and characterize human-made objects that orbit the Earth. Our efforts contribute to the collaborative, multiagency endeavor required to ensure comprehensive space situational awareness for the Nation.

* This is an edited, reformatted and augmented version of a statement presented March 19, 2013 before the House Committee on Science, Space, and Technology.

Air Force Space Command Roles and Responsibilities

Air Force Space Command presents space forces and capabilities to USSTRATCOM through the Fourteenth Air Force. The Commander, Fourteenth Air Force, Lieutenant General Susan Helms, is dual-hatted as the Commander, Joint Functional Component Command for Space (JFCC SPACE) and is responsible for executing USSTRATCOM's space operations mission.

JFCC SPACE's Joint Space Operations Center (JSpOC) is the avenue through which JFCC SPACE commands and controls space forces and it is the epicenter of the space situational awareness mission. The JSpOC is also the means by which JFCC SPACE coordinates space situational awareness with other agencies. For example, National Aeronautics and Space Administration (NASA) orbital safety analysts reside within the JSpOC 24 hours a day to collaborate on orbital safety threats to human space flight.

Detecting and Tracking Human-Made Objects in Space

For definitional purposes, the Air Force considers an object as being near-Earth if it takes less than 225 minutes to complete its orbit around the Earth. That is roughly 5,800 kilometers in altitude. All else is characterized as deep-space.

All entities that operate in the space domain are increasingly concerned about orbital debris. Past practices, as well as recent events, both accidental and purposeful, have created a troublesome debris environment in low Earth orbit. In 2007, the People's Republic of China performed an anti-satellite test which successfully struck its target, one of their defunct weather satellites. In 2009, an active Iridium communications satellite and a non-operative Russian Cosmos satellite accidentally collided. Each of these near-Earth incidents resulted in thousands of pieces of debris large enough to track. As of March 1, 2013, we continue to track 2,160 pieces from the Iridium-Cosmos collision alone. Additionally, in 2012, a Russian BRIZ-M upper stage malfunctioned with a significant quantity of propellant remaining. This upper stage eventually exploded and we are now tracking almost 150 pieces of debris from this event. More troubling is that our modeling tells us that each event produced thousands more pieces of debris which are too small for our sensors to reliably track. At orbital velocities, these small objects still represent catastrophic potential threats to fragile spacecraft. Each subsequent collision, explosion or break-up leaves more debris in space, increasing the potential for further collisions and even more debris, a chain reaction that could exponentially increase the risk to activities in space.

To support national security space operations in an environment of increasingly adverse environmental conditions, the JSpOC collects and processes data from a worldwide network of radar and optical sensors, as well as a dedicated space surveillance satellite. Each day the JSpOC creates and disseminates over 200,000 sensor taskings. The sensors then return nearly 500,000 observations to the JSpOC for processing. JSpOC operators use this data to maintain a high accuracy catalog of space objects and perform over 1,000 satellite collision avoidance screenings daily. These operations form the basis of the United States' space situational awareness capability, which is then shared with other operators in the national security, civil and commercial sectors of space operations.

SIZE OF OBJECTS AND DISTANCE DETECTED

The JSpOC directs space surveillance sensors to track objects in space ranging in size from as small as a softball to as large as the International Space Station. Today, the JSpOC tracks approximately 23,000 objects in both near-Earth and deep-space, but there is an estimated half million plus human-made objects in Earth orbit that we are not tracking. The number of objects reliably catalogued by the JSpOC is expected to rise by as much as four or five times by 2030 due to the steady growth in the on-orbit population, as well as the planned fielding of improved sensor capability.

The systems Air Force Space Command supports were designed and fielded to meet requirements specific to national security missions. Some track objects in near-Earth orbit while others are focused on deep-space, primarily geosynchronous orbit; however, we can on occasion support other orbit profiles. For example, when the NASA Stardust spacecraft returned to Earth from collecting samples of the coma of a comet in 2006, we were able to modify certain parameters of existing models and use space surveillance sensors to track the Stardust sample return capsule in its parabolic return-to-Earth.

TECHNOLOGIES AND PROCESSES

As previously stated, Air Force Space Command sensors were developed to track manmade objects in Earth orbit. The Nation's current capability to track asteroids, which orbit the sun, is largely driven by NASA. Air Force developmental telescopes are used by the Massachusetts Institute of Technology's Lincoln Laboratory to find and catalog asteroids under contract to NASA. And in some cases, the JSpOC can task space surveillance sensors to help track close approaches by asteroids and help predict potential collisions with Earth-orbiting objects. For example, during the recent Asteroid 2012 DA14 event, the JSpOC used orbit data for it from NASA's Near Earth Object Program Office at the Jet Propulsion Laboratory to screen for potential collisions with man-made objects in Earth orbit.

The current sensor tasking and data processing system used by the JSpOC to accomplish the space situational awareness mission was designed in the 1980s, fielded in the early 1990s and is nearing its capacity limits and end-of-life. We are in the process of fielding the next generation system, the JSpOC Mission System (JMS). With its open, service-oriented architecture, the JMS will supply the automation necessary to make better use of the tremendous volume of sensor data available. It will also enhance the Commander, JFCC SPACE capability to conduct space operations in a much more efficient and much safer manner.

CONCLUSION

Space situational awareness is foundational to civil, military and commercial space activities. Air Force Space Command forces and capabilities to detect, track, identify and characterize man-made objects in Earth orbit support the larger collaborative effort to maintain space situational awareness for the Nation.

In: Near-Earth Objects
Editor: Leon Sinclair
ISBN: 978-1-63484-067-5

Chapter 5

STATEMENT OF CHARLES BOLDEN, ADMINISTRATOR, NATIONAL AERONAUTICS AND SPACE ADMINISTRATION. HEARING ON "THREATS FROM SPACE: A REVIEW OF U.S. GOVERNMENT EFFORTS TO TRACK AND MITIGATE ASTEROIDS AND METEORS, PART 1"*

Mr. Chairman and Members of the Committee, thank you for the opportunity to appear today to discuss the topic of Near Earth-Objects (NEOs). NEOs are defined as those non-manmade objects in space whose orbits bring them within 1.3 Astronomical Units of the Sun, or to a set distance to the Earth's orbit that is generally equivalent to approximately 50 million kilometers (31 million miles).

The events of February 15, 2013 were a stark reminder of why NASA has for years devoted a great deal of attention to Near Earth Objects and why this hearing is so timely and important. The predicted close approach of a small asteroid, called 2012 DA14, and the unpredicted entry and explosion of a very small asteroid about 15 miles above Russia, have focused a great deal of public attention on the necessity of tracking asteroids and other NEOs and

* This is an edited, reformatted and augmented version of a statement presented March 19, 2013 before the House Committee on Science, Space, and Technology.

protecting our planet from them -- something this Committee and NASA have been working on for over 15 years. The events of February 15th also highlight the wisdom of the Congress, the Administration and NASA in enabling the human exploration of an asteroid.

The new public attention is not hard to understand. The coincidence of having these two very rare events, happening on the same day, along with the unfortunate injuries to over 1,000 people on the ground in Russia, made this a very big news story. However, we should remember that the probability of any sizable NEO impacting the Earth anytime in the next 100 years is extremely remote. The small fraction of objects we have discovered which do have the potential to impact the Earth are tagged by the Minor Planet Center as Potentially Hazardous Objects (or PHOs) and are subjected to further study and observation to assess just how hazardous they may be. Smaller objects, such as the recent impact in Russia, will always be difficult to detect and provide adequate warning; however, progress will be made over the next decade in this area as well.

To put these two recent events in context, very small objects enter the Earth's atmosphere all the time. The larger NEOs (those with a maximum physical dimension of more than a meter) are generally referred to as either asteroids or comets, while smaller objects are referred to as meteoroids. Current estimates are that on average about 100 tons of material in the form of dust grains and small meteoroids enter the Earth's atmosphere each day. Objects the size of a basketball arrive about once per day, and objects as large as a car arrive about once per week. Our Earth's atmosphere protects us from these small objects, so nearly all are destroyed before hitting the ground and generally pose no threat to life on Earth.

However, while objects the size of the one that exploded over Russia, which we have assessed as a rocky asteroid about 17 meters in diameter and weighing from 7,000 to 13,000 metric tons, enter the Earth's atmosphere very rarely on human timescales, they do have serious consequences. NASA has been at the forefront in leveraging our own resources, as well as interagency, international, academic and commercial partnerships, to study both these rare and more common NEO close approach events, and to expand our knowledge about NEOs and the potential threat they pose to the Earth. The 2010 National Space Policy specifically directs NASA to take a leadership role to "pursue capabilities, in cooperation with other departments, agencies, and commercial partners, to detect, track, catalog, and characterize near-Earth objects to reduce the risk of harm to humans from an unexpected impact on our planet and to identify potentially resource-rich planetary objects."

NASA leads the world in the detection and characterization of NEOs, and is responsible for the discovery of about 98 percent of all known NEOs.. Over 15-plus years of collecting data on NEOs has helped to shape the scientific consensus about these objects and the potential threat they pose to the Earth. NASA is leading a wide array of activities related to NEOs, including a long-standing ground-based observing campaign, focused flight missions to study both asteroids and comets, as well as conceptual studies and technology development to improve our ability to find NEOs. NASA uses radar techniques to better characterize the orbits, shapes, and sizes of observable NEOs, and funds research activities to better understand their composition and nature. NASA also funds the key reporting and dissemination infrastructure that allows for world-wide follow-up observations of NEOs as well as research related activities, including computer modeling, sample analysis and workshops to disseminate information about NEOs to the larger scientific and engineering community. Consistent with NASA's role as outlined in President's National Space Policy, NASA continues to collaborate with the Executive Office of the President and the Department of Defense on planning and exercises for responding to future hazardous NEOs.

NASA is also developing new vehicles and capabilities, including the Orion Multi-Purpose Crew Vehicle and the Space Launch System, which will enable human exploration of the solar system beyond low Earth orbit. As the President stated in his April 15, 2010, speech at the Kennedy Space Center, NASA's intention is to "[send] astronauts to an asteroid for the first time in history." NASA is working to accomplish this mission by 2025. This mission, and the vital precursor activities that will be necessary to ensure its success, should result in additional insight into the nature and composition of NEOs and will increase our capability to approach and interact with asteroids.

Detection-Related Activities

NASA was tasked by Congress in 1998 to catalog 90 percent of all the large NEOs (those of 1 kilometer or more in size) within 10 years; these would be large enough that should they strike Earth, it would result in a global catastrophe. NASA worked with a number of ground-based observatories and partners as part of our Spaceguard survey to reach that goal; NASA has now catalogued an estimated 95 percent of all NEOs over 1 km in size. None of these known large NEOs pose any threat of impact to the Earth anytime in the foreseeable future.

As shown in the next graphic, the cumulative discovery of Near-Earth Asteroids, the largest subset of NEOs, started picking up dramatically in 1998 with the start of NASA's Spaceguard search program and the number of known NEOs has grown from a few hundred to nearly 10,000 in just 15 years.

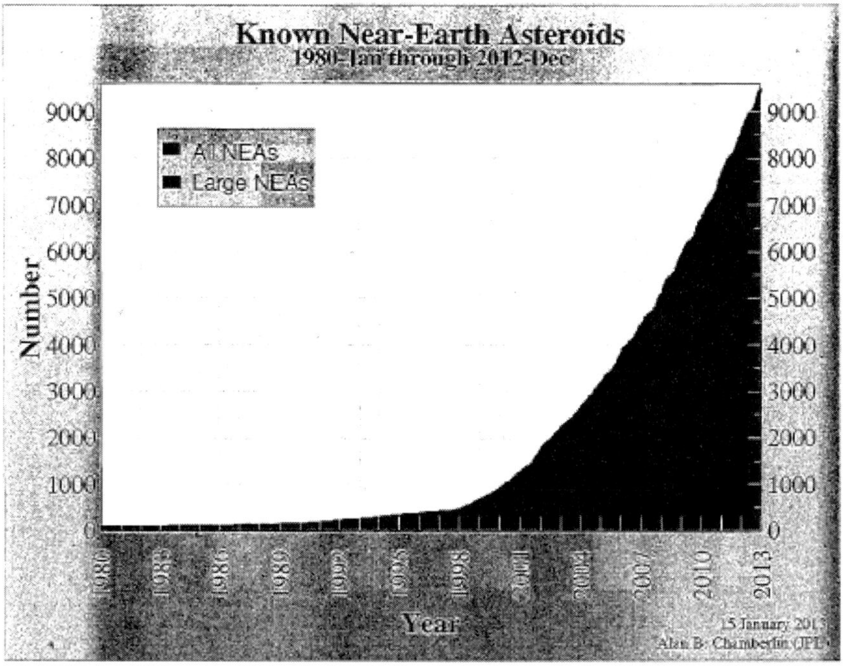

In 2005, as part of the NASA Authorization Act of 2005 (P.L. 109-155), Congress also directed NASA to initiate the George E. Brown, Jr. Near-Earth Object Survey, to "detect, track, catalogue, and characterize the physical characteristics of near-Earth objects equal to or greater than 140 meters in diameter" and set a goal for this program to achieve 90 percent completion by 2020. That effort is underway. NASA's NEO Observation Program currently funds three survey teams that operate five ground-based telescopes involved in the NEO search effort. Each team conducts independent operations for 14 to 20 nights per month, as weather permits, avoiding approximately a week on either side of the full moon when the sky is too bright to detect these extremely dim objects from the ground.

NASA also leveraged its investment in the Wide-field Infrared Survey Explorer (WISE) spacecraft to discover NEOs. WISE was designed originally as an astrophysics mission that would scan the entire sky in infrared light to study the coolest stars and the universe's most luminous galaxies. After completion of its astrophysics mission, WISE's operations were extended specifically as a NEO-finding mission. NASA made investments, referred to collectively as NEO-WISE, to adapt existing software, add data pipelines, and to put in place data collection, archive and retrieval software for the new NEO-WISE database. Overall, NASA's investments, in NEO-WISE resulted in the discovery of 146 previously unknown objects, including 129 near-Earth asteroids, of which 21 are Potentially Hazardous Objects, and 17 comets.

As noted in the following graphic, NASA continues to make progress toward the goals of the George E. Brown, Jr. Near-Earth Object Survey. To date, over 9,600 Near-Earth Asteroids of all sizes have been found. In addition to the 95 percent of NEOs over 1 km in size, our current estimate is that we have also found about 60 percent of the NEOs that are between 300 meters and 1 km.

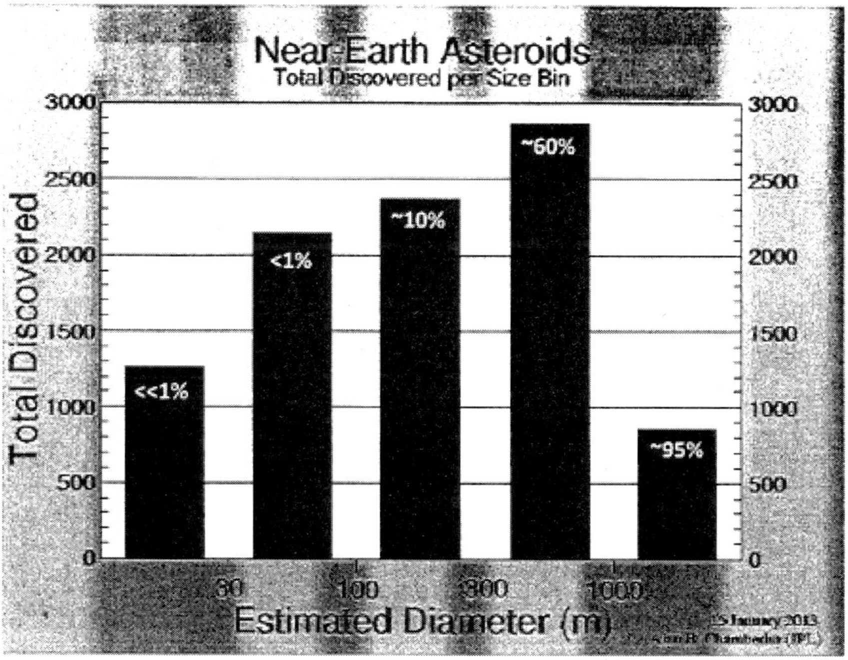

Risk Assessment and Characterization

Observations from all observatories around the world are sent to the international "clearinghouse" for small body observation data, the Minor Planet Center (MPC). The MPC maintains the database of observations and orbits on all known small bodies (asteroids, comets, dwarf planets, Kuiper Belt Objects, etc.) in the Solar System. It is funded by NASA and hosted by the Smithsonian Astrophysical Observatory's Center for Astrophysics. The MPC verifies and validates the observations by determining if they are of an already known object (by comparing them to the known orbits), or are indeed a new discovery. The MPC then determines and publishes an initial orbit for the new discovery so that observatories world-wide may look for the object and confirm its existence.

Once a new object's orbit is secured, its potential for impacting the Earth is assessed. As mentioned previously, well over 99 percent of all objects discovered have no potential for Earth impact even over many thousands of years, but the small fraction which do have some potential are tagged as PHOs by the Minor Planet Center. More detailed and refined analyses of a PHO's orbit, and an assessment of the risk posed by a particular object, are conducted by NASA's NEO Program Office at the Jet Propulsion Laboratory (JPL). Observations on PHOs are automatically forwarded to JPL and their orbits updated with high-precision analysis to determine a level of probability of the object impacting the Earth in the next 100 to 200 years. The results of this analysis are routinely updated and published on the NEO Program website at http://neo.jpl.nasa.gov.

When an NEO passes close enough to the Earth to be scanned by ground-based radar, NASA funds targeted radar observations. When an object passes close enough to the Earth to achieve a measurable radar return (about 20 million miles depending on the size), this allows NASA to obtain additional information about these objects. As was the case with 2012 DA14, the primary facility currently being used by NASA for routine planetary radar is NASA's own Goldstone facility, part of our Deep Space Network (DSN). NASA also uses the National Science Foundation's (NSF) Arecibo Radio Telescope facility for planetary radar observations. There are significant differences with the planetary radar capability at Arecibo compared to Goldstone. The Goldstone radar is a 70-meter steerable dish, allowing it to access objects significantly lower to the horizon than the more limited sky area accessible to the Arecibo radar. However, Arecibo is twice as powerful as Goldstone and has a much larger (304 meter) collection dish, which allows it to observe objects significantly farther away than Goldstone. Each plays an important role in the quick refinement of the orbit to a precision not obtainable by other means, and for understanding the object's size, shape and rotation rate. They can also aid in the detection of possible binary objects (-15 percent of NE0s), which in turn provides data that can be used to determine their mass.

NASA's NEO Observation (NEOO) Program has initiated development of several additional capabilities to the NEO detection network, with the recent additional funding it received starting in FY2012. Some of these involve collaboration on projects with the Defense Advanced Research Projects Agency (DARPA) and the U.S. Air Force, such as background detection of asteroids by the new Space Surveillance Telescope (SST), which is on track to start routinely providing observations this year. There is also the planned augmentation of the Panoramic Survey Telescope & Rapid Response System or Pan-STARRS facility with a second aperture. The wide field of view survey capabilities of these two assets are expected to provide a significant increase in NEO detection rate.

An important new development project that was started last year is the Asteroid Terrestrial-impact Last Alert System Project, or ATLAS Project. NASA's NEOO program has funded a five-year grant to the University of Hawaii to develop this innovative system. It will couple modest-sized, commercially-available telescopes with custom charge-coupled device cameras and rapid sky survey software to cover the entire available sky each night, detecting any asteroid bright enough to be seen by its detectors. ATLAS will find 100-meter sized asteroids millions of miles away, and smaller objects as they come closer to the Earth. It could provide days to weeks of warning of an object tens of meters in size that is on an impact trajectory with Earth. The prototype system is planned to begin sky testing by the end of calendar year 2014.

However, ATLAS, like SST and Pan-STARRS, will still be limited to the night sky and by weather. The only way to overcome these impediments is to use the vantage point of space. This is the idea behind a privately funded effort by the B612 Foundation to build a space observatory called Sentinel. NASA is providing B612 technical and operations assistance through a Space Act Agreement. Sentinel is being designed to find 100-meter sized objects and larger that could come near Earth's orbit.

To find the more numerous smaller asteroids near Earth, NASA also is investigating development of an instrument that could be hosted on geo-synchronous platforms such as communications, TV broadcast or weather satellites. This instrument would be a modest-sized, wide field telescope with detectors that operate in the infrared bands where these faint asteroids are more easily detected against the cold background of space. Though limited by the telescope size that can be hosted by a commercial geosatellite, such a capability shows promise to increase the detection rate of near Earth asteroids. We plan to initiate this project with an instrument solicitation later this year. The NASA Science Mission Directorate is testing detectors initiated in its Discovery Program that may be used in such a hosted asteroid detection telescope.

And, to further leverage international capabilities, NASA has been a leading participant in the NEO activities of the United Nations Committee on the Peaceful Uses of Outer Space (UNCOPUOS). Over the past several years, a working group on NEOs under the UNCOPUOS Scientific and Technical Subcommittee has been examining the topic of Earth-threatening NEOs. Results of that work led to recommendations this year, endorsed by the Subcommittee, to broaden and strengthen the international network to detect and characterize NEOs, and to call for relevant national space agencies to form a group focused on designing reference missions for a NEO deflection campaign. NASA has been at the forefront of these activities and will continue to take on that role.

Reconnaissance Activities

NASA's remote ground-based observations of comets and asteroids have been augmented by close-up reconnaissance data from the agency's science missions. Since the main asteroid belt located between the orbits of Mars and Jupiter is the source of NEOs, it is important to study it to fully understand these objects. NASA's Galileo spacecraft flew by and studied the main-belt asteroids Gaspra in 1991 and Ida in 1993 on its way to Jupiter. NASA's Near Earth Asteroid Rendezvous or NEAR mission (later renamed NEAR Shoemaker) flew by the main-belt asteroid 253 Mathilde in 1997, and beginning in 1988 flew by, then orbited, and in 2001 landed on the near-Earth asteroid 433 Eros. NASA is cooperating with the European Space Agency on its Rosetta mission, which encountered the main-belt asteroids 2867 Steins in 2008 and 21 Lutetia in 2010, and is scheduled to rendezvous and land on the Near-Earth comet 67P/Churyumov—Gerasimenko next year. NASA's Deep Space-1 spacecraft flew by a small Mars crossing asteroid, 9969 Braille, in 1999. NASA's Stardust spacecraft flew by the main-belt asteroid 5535 Annefrank in 2002. NASA's Deep Impact mission was specifically designed to impact and observe the effects on a comet. After imaging the comet Temple 1 in July 2005, the larger "flyby" spacecraft pointed high-precision tracking telescopes at the comet and released the "impactor" spacecraft into the comet's path for a planned collision. NASA's Stardust spacecraft later flew by Tempel 1 in February 2011, to further study what happened after the impact. NASA collaborated on Japan's Hayabusa mission which in 2010 successfully returned a small amount of samples from its 2006 encounter with the near-Earth asteroid 25143 Itokawa; 10 percent of these microscopic samples will be available for research by U.S. scientists as a result of NASA's agreement with the Japan Aerospace Exploration Agency (JAXA).

Most recently, NASA's Dawn spacecraft spent a year orbiting and observing the large main-belt asteroid 4 Vesta. Dawn provided close-up views of Vesta and unprecedented detail about the giant asteroid. Dawn's observations confirmed that the asteroid had completely melted in the past, forming a layered body with an iron core. The spacecraft also revealed the collisions Vesta suffered in its southern hemisphere. The asteroid survived two colossal impacts in the last 2 billion years. Without Dawn, scientists would not have known about the dramatic troughs sculpted around Vesta, which are ripples from the two south polar impacts. Dawn departed Vesta last August, and is now on its way to its planned rendezvous in 2015 with Ceres, the only dwarf planet in the inner solar system and the largest asteroid, so large that it is estimated to contain a third of the mass of the entire main asteroid belt.

These reconnaissance missions have greatly redefined what we know about asteroids and comets. These missions have observed asteroids that are binary systems, like Ida and Dactyl, where two objects travel together through space orbiting each other, which provides us insight into how these objects are formed and provide challenges for getting close to them. They have visited asteroids with primitive compositions that teach us about the origin of our solar system, and others that have undergone many of the same processes as terrestrial planets, providing a mirror for understanding our own. For example, data from our NEAR/Shoemaker mission suggests that Eros is a cracked but solid rock, probably a fractured chip off a larger body, made of some of the most primitive materials in the solar system. Judging from the meteorites, we have many more interesting varieties to visit in the future, from asteroids made of solid iron nickel alloys, to cosmic rubble piles, having collected bits and pieces over time and are only loosely connected.

A significant part of NASA's exploration of asteroids is to characterize the variety of asteroid composition and other characteristics that have both fundamental science value and provide critical information to inform the potential risk of PHOs. To this end, NASA is moving toward confirmation later this year for the Agency's planned asteroid rendezvous and sample return mission, dubbed OSIRISREx (for Origins-Spectral Interpretation-Resource Identification-Security-Regolith Explorer) that is planned to launch in 2016. After traveling three years, OSIRIS-REx will approach a near Earth asteroid, currently named 1999 RQ36, in 2019. 1999 RQ36 is believed to be the most exciting, accessible, volatile and organic-rich remnant currently known from the early Solar System; it also is thought to have abundant regolith (a blanket of loose materials covering rock), comprised of fine gravel that is ideal for collecting a sizable sample. With this sample, scientists will be able to analyze the asteroid's composition, mineralogy and geology to learn more about it and other organic-rich B-type asteroids. Once within three miles of the asteroid, the spacecraft will begin six months of comprehensive surface mapping. The science team then will pick a location from where the spacecraft's arm will take a sample of between 60 and 1000 grams (up to 2.2 lbs) for return to Earth in 2023. The careful study of 1999 RQ36 will permit scientists to fully understand the context in which the sample was selected, which should greatly increase the scientific value of the sample. NASA recently sponsored a contest for students worldwide to re-name 1999 RQ36 and draw attention to the important issues surrounding NEOs. The contest deadline was December 2, 2012, and we anticipate announcing a winner in the coming months.

NASA is also in discussions with our international partners to collaborate on several missions or mission concepts that could, in the future, grant access to U.S. researchers to valuable data on asteroids. NASA is working with JAXA on potential collaboration on the Japanese-led Hayabusa II mission. NASA is also discussing with the European Space Agency potential collaboration on two of their mission concepts: 1) the Marco-Polo-R mission concept which is focused on returning a sample from a primitive near-Earth asteroid in the late 2020s, and, 2) the Asteroid Impact and Deflection Assessment (AIDA) mission concept that could be used to study the binary asteroid system Didymos with two spacecraft and see if a small interceptor can affect any the change in the relative orbit of the two bodies.

Finally, NASA is working to accomplish an astronaut visit to an asteroid by 2025. This mission, and the vital precursor activities that will be necessary to ensure its success, should result in additional insight into the nature and composition of NEOs and will increase our capability to approach and interact with asteroids.

CONCLUSION

NASA has a long history of observing comets and asteroids but as their importance has become apparent as potentially hazardous objects, NASA has significantly increased its program of detection, reconnaissance, and characterization. NASA's current program utilizes extensive ground-based telescope observations in partnership with academia, U.S. Air Force, the National Science Foundation, and many international groups as well. We have a nearly complete understanding of our largest NEO population. NASA has determined that it is unlikely that the world will suffer a global catastrophic impact over the next several hundred years similar to the dinosaur extinction event. We are making marked progress in assessing the risk to our planet from smaller objects that could produce regional disasters. NASA is regularly reevaluating the risk to our planet and constantly updating our knowledge of the NEO population. Smaller objects, such as the recent impact in Russia, will always be difficult to detect and provide adequate warning; however, progress will be made over the next decade in this area as well.

Again, thank you for the opportunity to testify today, and I look forward to responding to any questions you may have.

In: Near-Earth Objects
Editor: Leon Sinclair
ISBN: 978-1-63484-067-5

Chapter 6

HEARING CHARTER FOR "THREATS FROM SPACE: A REVIEW OF PRIVATE SECTOR EFFORTS TO TRACK AND MITIGATE ASTEROIDS AND METEORS, PART 2"*

House Committee on Science, Space and Technology OR Committee on Science, Space and Technology

PURPOSE

At 2:00 p.m. on April 10, 2013, the Committee on Science, Space, and Technology will hold a hearing titled *Threats from Space, Part II: A Review of Private Sector Efforts to Track and Mitigate Asteroids and Meteors*. This is the second hearing this Congress where the Committee examines the tracking, characterization and mitigation of Near Earth Objects. The hearing will focus on the most viable near-term initiatives within the private sector and the international coordination needed to identify and characterize potentially hazardous near Earth objects.

---

* This is an edited, reformatted and augmented version of a document presented April 10, 2013 for a hearing conducted by the House Committee on Science, Space, and Technology.

WITNESSES

- Dr. Ed Lu, Chairman & CEO, B612 Foundation
- Dr. Donald K. Yeomans, Manager, Near-Earth Objects Program Office, Jet Propulsion Laboratory
- Dr. Michael F. A'Hearn, Vice-Chair, Committee to Review Near-Earth Object Surveys and Hazard Mitigation Strategies, National Research Council

OVERVIEW

On Friday, February 15, 2013, two events occurred that received worldwide attention. An unforeseen meteor (estimated 50 feet in diameter) exploded in the sky above the Russian city of Chelyabinsk releasing the equivalent of a 300 kiloton bomb, about twenty times the explosive energy of the atomic blast used over the city of Hiroshima. This blast injured nearly 1,200 people and resulted in an estimated $33 million in property damage. On the same day, a small asteroid (150 feet in diameter) discovered by amateur astronomers and tracked closely by NASA passed safely by the Earth, but within the orbital belt of geostationary satellites. Until it entered our atmosphere, the Russian meteor went completely undetected. According to NASA, the two events were unrelated, but raised public awareness of the potential threat from Near Earth Objects (NEOs).

On March 19, 2013, the Committee on Science, Space, and Technology held a hearing titled "Threats from Space: A Review of U.S. Government Efforts to Track and Mitigate Asteroid and Meteors." That hearing addressed the U.S. government's plans and programs to identify, catalog, and coordinate the threat of NEOs.

This hearing will look beyond just the U.S. Government to hear about public-private partnership, commercial private sector, and philanthropic initiatives to survey the sky for asteroids and comets.

Some of the overarching questions:

- Do we have the tools and technology necessary to detect and track Near Earth Objects?

- What are the most viable efforts to focus on in the next 5 to 7 years that will yield the most progress in identifying and cataloging the NEO threat?
- Are we tracking the right size objects, specifically the ones that can cause significant harm?
- Once we identify an object, what are our means of tracking it?
- What are our contingencies and mitigation capabilities if we determine there is a threat from a NEO impact?
- The Science, Space, and Technology Committee has been on the forefront of the issues surrounding Near Earth Objects. For example, the NASA Authorization Acts of 2000 and 2005 directed NASA to conduct a survey of the population of NEOs and study mitigation plans. NASA estimates 20,000 potentially hazardous asteroids orbit within the vicinity of the Earth.

In: Near-Earth Objects
Editor: Leon Sinclair
ISBN: 978-1-63484-067-5

Chapter 7

STATEMENT OF DR. EDWARD T. LU, CEO, B612 FOUNDATION. HEARING ON "THREATS FROM SPACE: A REVIEW OF PRIVATE SECTOR EFFORTS TO TRACK AND MITIGATE ASTEROIDS AND METEORS, PART 2"*

My name is Ed Lu, and I am the CEO of the B612 Foundation. Thank you for the opportunity to testify before the House Science, Space and Technology Committee to describe the B612 Foundation Sentinel Space Telescope project and its importance. The B612 Foundation is a Silicon Valley based nonprofit that is building, launching, and operating the Sentinel Space Telescope, which will find and track threatening asteroids.

NASA's Spaceguard Survey has already discovered more than 90 percent of asteroids larger than 1km. Why must we identify and track asteroids smaller than this?

The impact of a 1 km or larger asteroid would have energy of 40 thousand megatons, and would likely end human civilization regardless of where on the Earth it occurs. However smaller yet still potentially catastrophic asteroids are still largely not tracked. For example, the impact of a 140 meter asteroid would release several times more energy than all the munitions used in WWII. Yet have only observed and tracked less than 10% of asteroids in this size

* This is an edited, reformatted and augmented version of a statement presented April 10, 2013 before the House Committee on Science, Space, and Technology.

range. The impact of a 40 meter asteroid "city killer" such as the one that struck Tunguska on June 30, 1908 obliterated nearly 1000 square miles. Yet we have only observed and tracked less than 1% of asteroids in this size range. We have the technology to deflect asteroids to prevent an impact on Earth, but this technology is useless until we find asteroids first. We cannot deflect (or for that matter capture, visit, or explore) an asteroid that we haven't yet found. We simply do not know when the next catastrophic asteroid impact will be, because we have not yet tracked the great majority of asteroids.

Why is an infrared space telescope needed to discover and track these smaller yet still dangerous asteroids?

Asteroids are not only small but are often as dark as charcoal. That makes asteroids difficult to spot with ground based optical telescopes because not only are they dim, but the background sky is bright. But the fact that asteroids are dark can be used to our advantage if we observe in infrared. Asteroids are much easier to detect in infrared because they are warmer and thus brighter than the background sky in these wavelengths, and can therefore be seen at much greater range using infrared as opposed to optical. Such infrared observations can only be made from space, as the Earth's atmosphere absorbs these infrared wavelengths of light. The National Academies Report "Defending Planet Earth" published in 2010 describes how finding a substantial fraction of "city killer" asteroids like the 1908 Tunguska asteroid will require a space based infrared telescope.

Finding and tracking the roughly 1 million asteroids of this size in a reasonable timeframe requires a system capable of finding tens to hundreds of thousands of asteroids per year. This cannot be done even by large ground based optical telescopes, and it especially cannot be done by small optical telescopes. That means amateur astronomers unfortunately will not substantially contribute to this effort, and neither will small space based optical telescopes which some commercial companies have proposed to operate. Such a task requires an infrared space telescope.

Why is it critical to place Sentinel in Solar orbit (similar to the planet Venus)?

Asteroids that will hit Earth have orbits that cross Earth's orbit, and therefore are sometimes located in the direction of the Sun when viewed from Earth. Earth based telescopes cannot observe these asteroids when they are located in their "blind spot", i.e. when they are interior to the Earth. However, Sentinel will orbit the Sun interior to the Earth, in a solar orbit similar to that of the planet Venus. From that vantage point, Sentinel will be able to continuously look outwards away from the Sun while scanning Earth's orbit. This vantage point combined with Sentinel's ability to track asteroids from greater distances, means that Sentinel will typically be able to track an individual asteroid for several months at a time, which allows the orbit of that asteroid to be determined accurately. This is critical because many asteroids will have orbits which at first may appear to pose a threat to Earth until further observations can be used to refine our knowledge of the asteroid orbit well enough to rule out an impact. This is problem for telescopes located on or near Earth, as many asteroids can only be observed for a few weeks and then cannot be observed for long periods of time (often many years) because these asteroids recede in their orbits to the other side of the Sun for extended periods. Sentinel will orbit the Sun every 8 months, and so it will be able to observed and track these asteroids much more frequently, and therefore will be able to refine the orbits of such asteroids much faster. This will reduce incidences of asteroids having long periods of uncertainty such as we witnessed for the asteroid Apophis from 2004 until about 2010 (when our data was insufficient to be able to rule out an impact with Earth).

What is the role of the B612 Foundation and what makes it unique?

The B612 Foundation is a Silicon Valley based nonprofit that is philanthropically funded. We are funding, building, launching and operating the Sentinel Space Telescope. Because the problem of asteroid impacts is inherently a worldwide problem, the B612 Foundation has donors and supporters from around the world.

Our prime contractor is Ball Aerospace, located in Boulder, CO. Ball has previously built the Kepler Space Telescope, and the Spitzer Infrared Space Telescope on which Sentinel is largely based. We do have some non-financial support from NASA, which is providing use of the antennas of the Deep Space Network for telemetry and tracking, in addition to some technical consulting.

One of the unique aspects of the Sentinel mission is the way it is being managed. We are procuring Sentinel under commercial fixed price terms, much like communications and Earth observing spacecraft are procured. This is the first interplanetary mission to be managed in this fashion. We believe this is possible because Ball Aerospace has substantial experience with similar missions and much of Sentinel is based on that hardware, because there are no fundamentally new technologies which must be developed, and because both B612 and Ball have assembled world class technical teams. The B612 Foundation is managing this project in an innovative Silicon Valley fashion but with the rigor of a NASA project.

What is the status of the Sentinel Space Telescope project?

Sentinel is planned to launch in July 2018. The technical and management teams at both B612 and Ball are largely in place. There are 8 major milestones between project inception and launch. The first major milestone, the Concept and Implementation Review, was completed in September of 2012. The next major milestone, the Systems Definition Review, is scheduled for late fall of 2013. Prototype infrared imaging detectors have been fabricated and are currently in test.

In: Near-Earth Objects
Editor: Leon Sinclair

ISBN: 978-1-63484-067-5

Chapter 8

STATEMENT OF DONALD K. YEOMANS, MANAGER, NEAR-EARTH OBJECTS PROGRAM OFFICE, JET PROPULSION LABORATORY. HEARING ON "THREATS FROM SPACE: A REVIEW OF PRIVATE SECTOR EFFORTS TO TRACK AND MITIGATE ASTEROIDS AND METEORS, PART 2"*

Mr. Chairman and members of the Committee, thank you for the opportunity to appear today to discuss the progress and plans being made to discover, track and characterize the population of near-Earth objects that may pose threats to Earth.

The Importance of Near-Earth Objects: Near-Earth objects, commonly called NEOs, are comets and asteroids that can pass within about 28 million miles of the Earth s orbit. While icy active comets may occasionally pass close to Earth, it is the difficult-to-find, but far more numerous asteroids that are of most concern in near-Earth space today. Near-Earth objects are scientifically important because they represent the bits and pieces left over from the solar system formation process. Collisions with the early Earth likely brought much of the water and carbon-based materials that were the building blocks of life.

* This is an edited, reformatted and augmented version of a statement presented April 10, 2013 before the House Committee on Science, Space, and Technology.

Once life did form, subsequent collisions punctuated evolution, allowing only the most adaptable species to evolve further. We humans likely owe our origins and current position atop the food chain to these near-Earth objects.

While the vast population of near-Earth objects is a relatively recent discovery, they are of utmost importance in the study of the solar system s origin and our own origins, and they will likely play a major role in the future, providing building materials, water and fuel resources for interplanetary exploration and development. It is ironic that the near-Earth objects that are the easiest to reach for robotic or human exploration, and the easiest to exploit for their mineral and material wealth, are the same objects that represent the most serious potential threats to Earth. While finding them is important for future space resource development, we also need to find them - before they find us.

A Recent Hit and a Miss: Less than two months ago, on Friday, February 15, 2013, a 60-foot sized asteroid, traveling at 42,000 miles per hour, entered the Earth s atmosphere near Chelyabinsk in central Russia. It was heated and violently compressed by the atmospheric pressure and exploded about 14 miles above the surface, producing a descending shock wave with an energy of approximately 440 thousand tons of TNT explosives. Most of the asteroid itself was reduced largely to dust, but it also produced thousands of small fragments which fell to the ground as meteorites. Over 1,200 people were injured by the effects of the shock wave mostly from broken glass. Given the millions of similarly sized objects in Earth s neighborhood, a collision by one would be expected about once every 100 years on average. Coincidently, only 16 hours later, a larger 130-foot asteroid called 2012 DA14 that we had been tracking for a year came from a different direction, passing within 17,200 miles of the Earth s surface, 5,000 miles within the ring of communications satellites broadcasting the news of its arrival. An approach this close to Earth s surface by an object of this size is expected to occur every 40 years or so, on average. So, on the same day, we witnessed a once in a 100-year event and an unrelated once in a 40-year event. Here we have a nice example for science teachers to show that even extremely unlikely events in nature do happen even two on the same day.

Because it was found a year in advance, we were able to accurately predict the close Earth passage of asteroid 2012 DA14 on February 15, and we knew that it would not hit the Earth. However, the small asteroid that impacted the Earth s atmosphere over Russia arrived unannounced because it came from the direction of the Sun, and was hence unobservable with Earth-based telescopes. Discovering and identifying relatively small Earth impactors among the millions of asteroids in the Earth s neighborhood represents a significant challenge. Because there are so many more smaller asteroids than larger ones, the smaller ones hit the Earth s atmosphere more frequently. There are about ten million 20-meter sized asteroids like the one that exploded over central Russia two months ago, and their frequency of collision with the Earth is about once every 100 years, on average. When these small ones do hit, we expect them to break up in the atmosphere and cause only localized damage on the ground. Asteroids larger than one kilometer, on the other hand, could not only penetrate the atmosphere and impact the Earth s surface, they could also cause ejecta clouds that can affect weather patterns, produce firestorms and acid rain, and seriously harm global society and economics. The number of such large asteroids, however, is far smaller, only about 1000, and the frequency with which they impact the Earth is much lower, only about once every 700,000 years on average. But, the events of February 15, 2013, demonstrate that even extremely improbable events can happen, and that it is prudent to pay attention to the problem of finding and tracking all potentially hazardous near-Earth asteroids. And the focus should not be restricted to just the large near-Earth asteroids; mid-sized objects in the 100 to 500 meter range also pose a serious risk, since they could devastate an entire regional area.

The Spaceguard Goal: The population of near-Earth objects one kilometer and larger appropriately received the most attention in the early years of NASA s Near-Earth Object Observations program. In May 1998, a NASA representative announced plans to the House Subcommittee on Space and Aeronautics: NASA would find and track at least 90% of all the near-Earth objects larger than one kilometer; this became known as the Spaceguard goal. In December 2005, President Bush signed into law the George E.

Brown, Jr. Near-Earth Object Survey Act, that gave NASA a broader and more ambitious goal, to detect and track at least 90% of the near-Earth objects larger than 140 meters in diameter, and to characterize the physical properties of a representative sample of this population.

Significant Progress Has Been Made: When I last had the honor to address this Committee in November 2007, about 80% of the NEOs one kilometer or larger had been discovered and only a few percent of the smaller 140 meter objects. Today, the Spaceguard goal of discovering 90% of the large NEOs has been exceeded and about 25% of the 140 meter or larger sized NEO population has been discovered. Today, the discovery rate of NEOs is about 1000 per year, up 50% since 2007. The Minor Planet Center in Cambridge, Massachusetts, has 100 million observations of NEOs in its database and 27,000 observations are added daily. Fully 96% of all NEOs were discovered by NASA-funded surveys. The vast majority of all current NEO discoveries are being made by the Catalina Sky Survey, operating near Tucson, Arizona, the PanSTARRS survey operation atop Haleakala, Maui, Hawaii, and the LINEAR survey near Socorro, New Mexico.

None of the NEOs found to date have more than a tiny chance of hitting Earth in the next century. Thus the near-term risk of an unwarned impact from large asteroids, and hence the majority of the risk from all NEOs, has been reduced by more than 90%. Assuming none are found to be an impact threat, discovering 90% of the 140 meter sized objects will further reduce the total risk to the 99% level. By finding these objects early enough and tracking their motions over the next 100 years, even those rare objects that might be found threatening could be deflected using existing technologies. For example, a spacecraft could purposely ram the asteroid, modifying its orbital velocity by a very small amount, so that over several years its trajectory would be modified and its predicted impact of Earth in the future avoided by a safe margin. The autonomous spacecraft navigation required to effect such a collision was successfully demonstrated in July 2005 when NASA s Deep Impact spacecraft purposely rammed comet Tempel 1 to better understand the comet s structure and composition.

There have also been dramatic increases in the rate with which observations have been made to understand the physical nature of these NEOs, their so-called characterization . These include infrared observations that are used to infer asteroid sizes and compositions and radar observations that are used to determine asteroid sizes, shapes, rotation rates and whether or not an asteroid has a moon. For example, since 2007 when I last addressed this Committee, there has been a 250% increase in the number of infrared observations of NEOs made at the NASA supported Infrared Telescope Facility in Hawaii. During 2012 alone, the number of radar detections of NEOs at both the Goldstone facility in California and the Arecibo facility in Puerto Rico has tripled compared to the average of the previous ten years. Largely as a result of NASA support, there has been extraordinary progress in the last six years for the discovery and physical characterization of NEOs.

There has also been significant progress within the NEO Action Team associated with the UN Committee on the Peaceful Uses of Outer Space (COPUOS) to encourage and integrate more international efforts for NEO detections, for addressing deflection issues and for establishing protocols to be used by the international community in response to a potential NEO threat.

Possible Next Steps For Ground-Based Systems: There is still much work to be done. About 50-100 NEOs larger than one kilometer remain undiscovered, along with about 13,000 NEOs larger than 140 meters and millions of objects larger than about 30 meters in extent - the approximate minimum size for a common stony asteroid to cause significant ground damage.

It is important that the current NEO discovery surveys, operating with one to two meter-class optical telescopes continue their nightly searches and continue to improve their equipment, operations and data processing. These ground-based optical telescopes will continue to significantly add to the count of large NEOs discovered, but the current search assets will not be able to reach the goal of finding nearly all of the population of 140-meter sized objects within a reasonable time period because of their limited capability to detect these very dim objects.

The existing Pan-STARRS1 (PS1) system operates a 1.8-meter aperture telescope on the island of Maui but this instrument only focuses its attention on NEO observations for about 11% of its observing time because of other science objectives. Even so, PS1 currently provides about 25% of the NEO discoveries, second only to the Catalina Sky Survey. Suitable funding to increase the percentage of time devoted to NEO searches on Pan-STARRS1, at the expense of other science, would accelerate the current NEO discovery rate, as would the full time or part time use of a second Pan-STARRS2 telescope that is nearing completion adjacent to the Pan-STARRS1 facility on Maui.

An important planned future contributor is the Space Surveillance Telescope (SST), a 3.5-meter wide-field telescope that is being developed by MIT s Lincoln Laboratory for DARPA and the US Air Force. When fully operational in late 2014, this telescope will scan a wide region centered on the equatorial band of the night-time sky. Investigations are ongoing to better understand the efficiency with which this telescope will discover NEOs and what sort of scheduling might be intermingled with its prime mission of man-made space object surveillance to carry out these NEO observations.

The most effective, ground-based NEO detection telescope that is currently in planned development is the Large Synoptic Survey Telescope (LSST), a 8.4-meter aperture, wide-field telescope that is planned to begin operations in Chile in the early 2020s. To be funded by the National Science Foundation and a consortia of private and international agencies and universities for a variety of science programs, simulations have suggested that the shared use of LSST could catalog approximately 25% of the 140 meter sized NEOs within 5 years of operations and about 45% in ten years.

The View From Space: Especially for the population of undiscovered sub-kilometer sized objects, space-based infrared telescopes would be a more efficient discovery system than the current ground-based optical surveys. This is because asteroids emit considerable heat, not just reflected sunlight, and this heat makes them bright in the infrared wavelengths, but these wavelengths are also unfortunately heavily filtered by the Earth s atmosphere. In addition, the view from an observatory orbiting the Sun interior to the Earth s orbit would have far better viewing coverage of hazardous objects farther away from Earth. Furthermore, a space-based telescope would not have to deal with downtime due to weather and daylight. Ground-based telescopes have difficulty distinguishing a large, dark asteroid from a small, bright asteroid, often making asteroid size measurements very uncertain. On the other hand, space-based infrared measurements can infer an asteroid s size with an uncertainty of only about 10% and its reflectivity to about 20%. These types of measurements were demonstrated in 2010 when the highly successful NEOWISE effort mined asteroid discoveries, sizes and reflectivities in the data produced by the Wide-field Infrared Survey Explorer (WISE) satellite.

If the goal is to complete the survey of 140 meter sized objects more quickly, the 2010 National Research Council report entitled Defending Planet Earth indicated that a space-based infrared telescope in either a Venus-like orbit or interior to the Earth on the Sun-Earth line (L1 point) would be far more efficient finding NEOs than would existing, or planned, ground-based optical surveys. For the more numerous population of smaller NEOs that can still do significant ground damage, an infrared telescope at L1 would be well positioned to find those smaller objects making close Earth approaches. A successful space-based IR survey telescope in a Venus-like orbit would be very effective in discovering NEOs further in advance and providing positional observations unavailable from Earth-based telescopes. Together these observations would allow a faster refinement of an asteroid s orbit so that impact predictions could also be updated more quickly. Hence these space-based observations might provide an early all clear and avoid otherwise unnecessary concern and unneeded deflection mission planning or initiation.

Threat Mitigation vs. Threat Warnings: For the millions of small NEOs, in the range of 30 to 50 meters, it would be extremely challenging to find the majority of this population far enough in advance to first determine which ones represent a threat and then deflect them safely away from Earth. And meeting such a challenge may not be cost effective. It may be sufficient to simply detect these small asteroids a few days or weeks prior to Earth impact so that appropriate warnings could be made and evacuations undertaken similar to hurricane emergencies in the unlikely case where populated areas of

Earth would be threatened. A warning of this type would also assure affected nations that the coming explosive blast would be a natural phenomena rather than a hostile act.

The NASA-funded ground-based ATLAS system currently under development at the University of Hawaii is a relatively low cost, wide-field telescopic survey designed to patrol the entire accessible night sky every night to provide suitable impact warnings for small asteroids on near-term Earth impacting trajectories. Simulations suggest that the ATLAS system, consisting of 3 to 4 sites worldwide, will find almost all objects larger than 30 meters coming at us from the night sky and provide a week s warning time. Current search programs are designed to find larger potentially hazardous objects well in advance of a predicted impact so that existing technologies (e.g., spacecraft rendezvous and impacts) could be employed to deflect the object out of harm s way. One of the issues with which policy makers will need to wrestle is where to draw the line as to the minimum NEO size that represents so large a threat as to require deflection attempts. Objects below that limit would then require only advance warning. Cost benefit studies would shed some light on this issue.

Summary: The NASA-supported NEO observations program is proceeding extremely well, and the rate with which NEOs are being discovered and physically characterized is increasing each year. There are viable options for accelerating the current NEO search efficiencies either using next-generation, ground-based optical surveys or the even more efficient space-based infrared surveys. The use of both ground-based and space-based assets would be the most effective option for quickly finding 90% of the NEO population larger than 140 meters. Robust future NEO search programs and the attendant physical characterization efforts could provide a large number of target bodies for scientific study, for future robotic and human exploration and for future resource development. These same surveys could also identify which of the discovered NEOs represent potential future threats and do so with enough time to either deflect the object, or warn of its arrival.

In: Near-Earth Objects
Editor: Leon Sinclair
ISBN: 978-1-63484-067-5

Chapter 9

TESTIMONY OF MICHAEL F. A'HEARN, VICE-CHAIR, COMMITTEE TO REVIEW NEAR-EARTH OBJECT SURVEYS AND HAZARD MITIGATION STRATEGIES, NATIONAL RESEARCH COUNCIL. HEARING ON "THREATS FROM SPACE: A REVIEW OF PRIVATE SECTOR EFFORTS TO TRACK AND MITIGATE ASTEROIDS AND METEORS, PART 2"*

Mr. Chairman and members of the Committee, thank you for the opportunity to appear today to discuss the potential threats of near-Earth objects (NEOs) in the context of the NRC report on this topic that was issued in 2010. I was the chairman of the mitigation sub-panel for the NRC report, but today I am not representing the NRC, nor NASA, nor the University of Maryland.

The NRC Study: As mandated by Congress in the Consolidated Appropriations Act, 2008, NASA commissioned the NRC to study Surveys for Near-Earth Objects and Hazard Mitigation Strategies. The Steering Committee

* This is an edited, reformatted and augmented version of testimony presented April 10, 2013 before the House Committee on Science, Space, and Technology.

was chaired by Dr. Irwin Shapiro of Harvard University and the two sub-panels, one for Surveys and Characterization and one for Mitigation Strategies, were chaired by Dr. Faith Vilas, then Director of the MMT Observatory in Arizona, and by myself, respectively. The committee had a wide variety of expertise, ranging over the entire scope of the impact hazard problem. Several public hearings were held, with testimony from numerous experts, some of whom were advocates of specific projects while others were experts in impact prediction and risk communication, and yet others were policy experts.

The committee concluded that the money being expended at that time on NEO surveys was inadequate to meet the congressional mandate of finding 90% of potential impactors larger than 140 m on any reasonable time scale. The committee did not make a specific recommendation on the forward path, but described forward paths for surveys and discovery as a function of how much money Congress wished to appropriate to "buy insurance" against an impact. The amount of money to be appropriated would directly affect the timeline. The committee also recommended initiating a search for potential impactors in the 50-140-m range. The committee noted that there are basically four approaches to mitigation – evacuation for the smallest impactors, slow push-pull techniques, such as the gravity tractor, for moderately sized impactors with long warning times, and then kinetic impactors and standoff nuclear explosions for successively larger impactors and/or shorter warning times. A research program to better understand these mitigation approaches was recommended. Actual mitigation experiments in space were suggested, provided sufficient funding was provided, and overall programs were described for three different levels of funding.

The committee's report, Defending Planet Earth – Near-Earth-Object Surveys and Hazard Mitigation Strategies, was released in 2010. The remainder of this testimony concerns the details of some of these recommendations, both as recommended by the NRC and including my personal perspectives on the issues.

Impactors <140 meters: At the time of the NRC report, results newly published at that time indicated that previous modeling of impacts, by scaling from nuclear explosions of known yield, were incorrect due to the rapid downward motion of an external impactor compared to a nuclear explosion, for which the source can be considered to be at a fixed altitude. These results, which are still neither refuted nor explicitly confirmed, show that substantial damage can be inflicted by objects that are even smaller than 50 meters in diameter. To be specific, the new calculations suggested that the Tunguska event, which in 1908 flattened every tree over roughly 2000 square km in

Siberia, was due to a body in the range of 30-50 meters diameter. Based on our knowledge of the size distribution of NEOs, that corresponds to an event that should occur roughly every century or two. For comparison, the best estimate of the Chelyabinsk meteor in February, which caused one building collapse and lots of broken windows with many people injured, is that it had a diameter of 15-20 meters, much smaller than any of the previous estimates of a hazardous size. The size of the Chelyabinsk meteor is better known than most since the trajectory has yielded a reliable velocity and the recovered samples can be used to infer the density of the body. Such an event should occur every several decades. Thus it is clear that objects much smaller than 140 meters are frequent and are capable of significant damage on Earth, although most of these impacts in the past went unnoticed because they occurred over the ocean or over very sparsely inhabited land areas. Detailed modeling of the effects of small impactors, say from Chelyabinsk-size to 140-m diameter, is a gap that should be filled, although most of the computer codes to tackle this problem accurately are under restricted access.

It is widely understood that small objects are much more abundant than large ones in nearly all the populations of the solar system, and specifically among the NEOs. Very roughly, a 14-m NEO is 1000 times more likely than a 140-m NEO. Thus the "next" significant impactor will most likely be closer in size to Chelyabinsk than to 140 meters. It therefore is important to plan for such an event, even if the hazard to life is small.

A key issue for the small impactors is that they are normally so faint prior to impact that we do not know how to detect them very far in advance. Many of them can only be discovered days to weeks before impact. Fortunately, this limitation coincides with the fact that the region of destruction by such an impactor is sufficiently small that evacuation (aka "duck and cover") is a realistic mitigation to minimize loss of life (but not property damage). You will hear about current efforts related to the ATLAS system from Dr. Yeomans and about the private venture to deploy the Sentinel system from Dr. Lu. All other things being equal, space-based systems offer a major advantage in principle, as long as the orbit is sunward from the Earth, such as at Earth's L1 Lagrange point, because it avoids the need for multiple sites on the ground. However, a cost benefit analysis must be undertaken that includes, limiting magnitude, wavelength range of operation, and the data processing approach. The ATLAS system alone is not sufficient for reliable detection because it consists of only two telescope systems, i.e., telescopes at only two sites, but it is designed to be sufficiently low in cost that other countries could realistically deploy similar systems, thus providing 24-hour coverage of both northern and southern hemispheres. The real issue then will be simply implementing the real-time coordination among the systems.

Programs at Various Funding Levels: The NRC report noted that any program dealing with NEO hazards as policy, as opposed to programs dealing with NEOs as scientific targets, should be considered as a form of insurance. The hazard is different from other terrestrial hazards, however, in that the insurance can be used to prevent damage rather than paying for restoration after damage. The question should be thought of, therefore, as a question of how much insurance the nation should by. The committee then described three different scenarios, depending on how much insurance was being bought, with rather arbitrary levels being chosen for the scenarios.

At a level of $10 million per year, the then operating survey programs could continue, as could a modest research program into issues related to the NEO hazard. This level would not meet the congressionally mandated George E. Brown survey to detect 90% of potential impactors larger than 140 meters in diameter.

I note that current spending in NASA's NEO program has increased to roughly $20 million per year, allowing some new initiatives such as the ATLAS program, operations of the PanSTARRS system (currently only one telescope but soon to be two telescopes), and research grants into mitigation related topics. Spending for the Large Synoptic Survey Telescope is not included in these totals – that telescope, if operated in NEO survey mode only, could meet the 140-meter goal relatively quickly.

At a level of $50 million per year, operation of a telescope such as LSST could be funded for NEO-optimized searches, although this assumes construction funding for astronomical research, *e.g.*, from NSF. Alternatively, an in-flight mitigation mission might be feasible if conducted as a minor part of an international partnership.

At a level of $250 million per year for a decade, the advanced surveys to 140 meters could be completed, either from the ground or from space, and a unilateral mitigation experimental mission would be feasible.

None of the NRC's recommended funding levels addressed the question of impactors smaller than 140 meters. With current technology, late detection appears to be the only feasible approach. Limits for the Sentinel system are not readily available to me, nor are the actual limits of the ATLAS system so I cannot comment on their relative contributions. The NEO program office at JPL has funded an independent study to assess the capability of the ATLAS system.

One also needs to remember that, once the George E. Brown survey to 140 meters is complete (90%), the remaining unidentified impactors include both the smaller impactors and the long-period comets. Although the long-period comets very rarely impact Earth, cumulatively they are likely to lead to as many or more deaths as the much more frequent small events. They have been ignored up to this point because they have been such a small fraction of the total threat, but that situation will change dramatically. One has to decide whether to deal with the small, frequent events or with the rare, large events, or both, analogous to deciding whether to deal with frequent auto accidents or infrequent large airliner or ship accidents or both.

International Cooperation and Collaboration: International collaboration is very important in the entire effort to deal with the impact hazard, from discovery, through impact prediction, to mitigation. Unfortunately, despite considerable discussion at the individual scientist level and considerable discussion at the governmental level up to the United Nations, the U. S. is the only nation with a funded, active and effective survey/discovery program. Canada has just launched (February 2013) and Germany will soon launch a small satellite designed to discover sub-populations of NEOs, but the U.S. is still the predominant nation in funding an active program for tracking NEOs, both through the JPL NEO Program Office and through funding the entire operation of the Minor Planet Center that is nominally sponsored by the International Astronomical Union.

It should be pointed out that the only terrestrial impactor ever predicted in advance was 2008 TC_3, an impactor much smaller (roughly 4 meters) than the Chelyabinsk meteor. This was discovered less than one day before impact, by R. Kowalski at the Catalina survey, based in Arizona. The impact was predicted only because the Catalina survey included a (NASA-funded) telescope in Australia in addition to the telescopes in Arizona, which allowed very rapid follow up data, and it was the combination of data from both telescopes that allowed the rapid prediction of the impact, including a prediction of the time and location of impact, both of which were extremely accurate. Thus an internationally distributed, and closely interactive, network of telescopes is critical for predicting small impactors. Fortunately, 2008 TC_3 was so small that it caused no damage on the Sahara Desert in northern Sudan where it entered Earth's atmosphere, although small pieces were subsequently recovered days later.

The area in which international collaboration is even more important is mitigation, due largely to the fact that incorrectly changing the orbit of a potential impactor could merely move the impact site from one country to another, with obvious international implications. Even the Chelyabinsk meteor was claimed by a fringe politician in Russia to be an American weapons test, but fortunately the Russian Academy of Sciences was in the forefront of public announcements, clearly declaring that this was a natural meteor. Unfortunately, there has been even less international discussion on this topic than on the survey/discovery/prediction topic, although there have been discussions within the UN's Action Team 14 of COPUOS. This is an area in which international collaboration, not just discussion, must be established before action is needed.

Contributions of Basic Research to Detection, Characterization, and Mitigation: There is considerable overlap between basic scientific research on comets and asteroids, *i.e.*, on the bodies that include NEOs, and policy-based work on the issues of hazard prediction and mitigation. However, the focus is very different between the two areas and consequently there are significant activities that are not included in one focus or the other. It is for this reason that NEO hazard activities require a separately identified source of funding, associated with national policy, that is not taken out of the scientific programs.

The research activities related to surveys and discovering bodies are aimed at finding statistically significant samples to enable interpretation, and these were the precursors of the specific hazard surveys, which are aimed at discovering as close to all of the objects as is practical (widely being taken to be 90% of the estimated total population). The research surveys, coupled with the work of dynamical researchers studying the orbits of the bodies, are what led to the recognition of the scale of the hazard and many of the individuals involved in those surveys are also involved in the hazard-driven surveys.

Research activities are also directly related to mitigation, but clearly distinct from actual mitigation planning. One of the key issues in mitigation, and for that matter even in predicting the scale of the damage from an impact, is to understand the physical properties of the impactors. Research programs using remote sensing have shown unambiguously that there is a wide variety of physical characteristics among the NEOs, ranging from likely coherent bodies that are the source of iron meteorites through really porous cometary nuclei that are likely to have been the source of the dinosaur-killer K-T impact 65 million years ago. Remote sensing can study a large number of objects and they are sensitive primarily to surface properties of the objects, to their size, and in some cases to a crude measure of their shape and their density.

Important, detailed characteristics of the NEOs can only be learned from *in situ* studies and PI-led, competitively selected missions, under NASA's Discovery and New Frontiers programs, provide the key mechanism to carry out these studies. Such missions can only be used to study a very few targets for budgetary reasons. A team led by Mike Belton and myself proposed the Deep Impact mission to the Discovery Program many years ago purely as a scientific mission, with only two sentences in the proposal about the possible peripheral benefits for NEO hazard mitigation. What the mission did for hazard mitigation was to demonstrate active targeting to impact on a small body, the nucleus of comet 9P/Tempel 1 (a technique needed for our science but also a technique needed for mitigation) and it also demonstrated the very porous nature of cometary nuclei (probably 10% of NEOs are inactive cometary nuclei). The observations of the ejecta were used both to determine the bulk density (much empty space inside!) and to estimate the momentum transfer efficiency of the impact as relatively low (roughly 2), a critical parameter for altering an NEO's orbit with a kinetic impactor. The mission also showed the challenges of attitude control in the last minute of approach to a cometary nucleus. These results have been presented to various groups directly concerned about mitigation, such as the Defense Threat Reduction Agency. The results of the subsequent flyby of comet Hartley 2 as part of the EPOXI mission showed the diversity among cometary nuclei and the heterogeneity from place to place on a single nucleus, both of which must be taken into account in mitigation.

The OSIRIS-REx mission, scheduled for launch in 2016, is a very different mission to a different type of NEO, the asteroid 1999 $_{RQ36.}$ This mission will return a sample of the asteroid to Earth for detailed analysis, but while at the asteroid it will also produce, for example, a detailed map of the gravity. In addition to the material properties learned from the returned sample, gravitational mapping can be used to understand the internal structure of the asteroid, critical information for understanding how to mitigate by changing the orbit, whether by kinetic impact, or nuclear explosion, or even with a gravity tractor, which depends less on the physical structure but does depend on the bulk density and the shape.

These competitively chosen "research" missions are not sufficient to completely address mitigation, but they provide most of the necessary information on the range of physical properties one might encounter. Unfortunately, the NASA budget for planetary exploration has been such that NASA's Discovery program (competitively selected, PI-led missions with a cost cap of $425M in the latest round), have been devastated compared to even a decade ago. The NRC's recent decadal survey of planetary science recommended that NASA's priorities should be first to maintain a cadre of good researchers, and then to maintain a regular cadence, averaging a new start every two years, for the smallest missions (the Discovery Program), then the New Frontiers program (similar to the Discovery Program but for missions twice as expensive), and finally flagship missions (center directed missions that have lately cost more than $2 billion). Although not every mission in Discovery and New Frontiers is relevant to hazard mitigation (the most recent selection in the Discovery program is a mission to Mars), restoring Discovery to the originally intended cadence of research missions would significantly help with the mitigation effort by ensuring the existence of other missions to comets and asteroids to provide information necessary for mitigation.

Ultimately, however, specific mitigation missions must be considered as discussed above under program levels. They should be funded over and above the research program and they could be either separately funded add-ons to scientific missions or stand-alone missions, or international collaborations, with the international collaboration a high priority. Note that once the range of physical properties is understood, it is still very difficult to determine the physical properties of an actually threatening NEO without sending a mission to it, a possibility with very early discoveries but not with late discoveries.

What Should be Done in the Event of an Identified NEO Threat? After an NEO threat is identified, the initial steps are well defined. NASA is the lead agency for identifying threats and they have a reporting path through the U.S. government that covers all relevant federal agencies and the POTUS. Reporting to other countries is also urgent and should be done through the U.N. in order to reach all governments. In addition, there should be direct communication with countries and international agencies that have relevant capabilities for mitigation. Immediately following the alert, it is crucial to share all available data publicly. This is routine for the positional observations of the NEO and for the resultant orbital computations through the Minor Planet Center and through JPL's NEO Program Office. Beyond this, however, it is crucial to share all available information on the physical characteristics of the NEO from whatever source and on the details of the impact prediction. In the case of 2008 TC_3, which presented no hazard, this information was communicated through the channels normally used worldwide by astronomers and information was made readily available to news media.

The next steps depend critically on the nature of the threat Ð how big the impact will be, how far in the future it will occur, and where it will occur. An all-out effort to determine the characteristics of the particular impactor is crucial Ð remote sensing being needed in any case and, if time permits, a mission to characterize the NEO should be initiated in order to optimize the mitigation. Short warning times, however, may preclude an advance characterization mission and in that case the range of expected properties must be used to design a fail-safe mitigation. Action paths are, to my limited knowledge, not yet in place domestically. For a small impactor, a plausible route is through FEMA. For a larger impactor, however, either the military or NASA might be the one to take charge. For truly large impactors, the lead country and agency should be coordinated among those countries that have the capability to execute any mitigation. This decision/action tree should be fleshed out and made publicly available long before any specific threat is identified.

INDEX

A

B

C